Memoir on the Geology of Central France: including the volcanic formations of Auvergne, the Velay, and the Vivarais.

George Poulett Scrope

Memoir on the Geology of Central France: including the volcanic formations of Auvergne, the Velay, and the Vivarais.
Scrope, George Poulett
British Library, Historical Print Editions
British Library
1858
8°.
7107.d.39.

The BiblioLife Network

This project was made possible in part by the BiblioLife Network (BLN), a project aimed at addressing some of the huge challenges facing book preservationists around the world. The BLN includes libraries, library networks, archives, subject matter experts, online communities and library service providers. We believe every book ever published should be available as a high-quality print reproduction; printed on- demand anywhere in the world. This insures the ongoing accessibility of the content and helps generate sustainable revenue for the libraries and organizations that work to preserve these important materials.

The following book is in the "public domain" and represents an authentic reproduction of the text as printed by the original publisher. While we have attempted to accurately maintain the integrity of the original work, there are sometimes problems with the original book or micro-film from which the books were digitized. This can result in minor errors in reproduction. Possible imperfections include missing and blurred pages, poor pictures, markings and other reproduction issues beyond our control. Because this work is culturally important, we have made it available as part of our commitment to protecting, preserving, and promoting the world's literature.

GUIDE TO FOLD-OUTS, MAPS and OVERSIZED IMAGES

In an online database, page images do not need to conform to the size restrictions found in a printed book. When converting these images back into a printed bound book, the page sizes are standardized in ways that maintain the detail of the original. For large images, such as fold-out maps, the original page image is split into two or more pages.

Guidelines used to determine the split of oversize pages:

- Some images are split vertically; large images require vertical and horizontal splits.
- For horizontal splits, the content is split left to right.
- For vertical splits, the content is split from top to bottom.
- For both vertical and horizontal splits, the image is processed from top left to bottom right.

[The map of Auvergne is missing from the first pocket. 7. iv. 1913.]

7107. d 39

THE GEOLOGY

AND

EXTINCT VOLCANOS

OF

CENTRAL FRANCE.

"Ce n'est pas avec d'immenses connoissances qu'on parvient toujours à mieux voir que les autres, c'est par une bonne méthode toujours secondée de l'observation. Surtout l'observation. C'est ce qui fait que le Chimiste, enfermé au milieu de son laboratoire et de ses fourneaux, est souvent un médiocre naturaliste et un mauvais géologue : il voit bien là, peut-être, comment la Nature a fait dans une pierre ; mais c'est dans les champs, c'est sur la cime des monts, qu'on voit comment elle a fait sur le globe."—MONTLOSIER, *Essai sur la Théorie des Volcans d'Auvergne*, p. 139.

"Il n'y a peut-être pas dans tout l'univers une contrée où les terrains volcaniques soient plus variés, mieux liés entre eux, et par conséquent plus instructifs, que dans le milieu de la France."—VON BUCH, *Lettre à Pictet*.

29 AP 58

THE GEOLOGY

AND

EXTINCT VOLCANOS

OF

CENTRAL FRANCE.

By G. POULETT SCROPE, M.P., F.R.S., F.G.S., &c.

Second Edition, enlarged and improved.

WITH ILLUSTRATIVE MAPS, VIEWS, AND PANORAMIC SKETCHES.

LONDON:
JOHN MURRAY, ALBEMARLE STREET.
1858.

LONDON: PRINTED BY WILLIAM CLOWES AND SONS, STAMFORD STREET,
AND CHARING CROSS.

PREFACE.

While resident in Italy during the winters of 1817-18-19, I had observed with great interest the volcanic phenomena of Vesuvius, Etna, and the Lipari Isles, and paid considerable attention to the structure of the district west of the Apennines, between Santa Fiora in Tuscany and the Bay of Naples, which presents unmistakeable traces of volcanic action on an extensive scale, though no eruption has taken place there within the historical period.

After my return to England, being for some time at Cambridge, I had the advantage of frequent intercourse with the late Professor E. D. Clarke, who was himself well acquainted with volcanic Italy, and Professor Sedgwick, at that time commencing his distinguished career as a geologist. The doctrines of Werner were then so completely in the ascendant that it was considered little better than heresy to dispute any of them. Yet it appeared to me, from the knowledge of igneous rocks I had acquired in Italy, that the dogmatic canon of that school which denied a volcanic origin to the Flœtz Trap-rocks (as basalt, clinkstone, and trachyte were then called), and declared them to be precipitations from some archaic ocean, was signally erroneous.

My two friends agreed with me in the opinion that the error of the Wernerians in undervaluing, or rather despising altogether as of no appreciable value, the influence of volcanic forces in the production of the rocks that compose the surface of the globe, formed a fatal bar to the progress of sound geological science, which it was above all things desirable to remove.

Being shortly after free to choose my path of travel, I determined to examine with care such evidence upon this point as might probably be found in Auvergne and the neighbouring districts—a country where the products of extinct volcanos are brought into contact with some of the earliest crystalline rocks, as well as with the most recent (tertiary and freshwater) strata.

For this purpose, in the beginning of June, 1821, I established myself at Clermont, the capital of the department of the Puy de Dôme, and passed some months in continual examination of the geology of the neighbourhood; removing from thence, as it became convenient, to the Baths of Mont Dore, Le Puy (Haute Loire), and Aubenas (Ardèche). I afterwards revisited Italy, where I had the good fortune to witness by far the most important eruption of Vesuvius that has occurred within this century—that of October, 1822.

PREFACE.

On my return to England in 1823, I published a volume on the 'Phenomena of Volcanos.'* In that work unfortunately were included some speculations on theoretic cosmogony, which the public mind was not at the time prepared to entertain. Nor was this, my first attempt at authorship, sufficiently well composed, arranged, or even printed, to secure a fair appreciation for the really sound and, I believe, original views on many points of geological interest, which it contained. I ought, no doubt, to have begun with a description of the striking facts which I was prepared to produce from the volcanic regions of Central France and Italy, in order to pave the way for a favourable reception, or even for a fair hearing, of the theoretic views I had been led from those observations to form.

Indeed this obvious error was pointed out in a very friendly manner by the Quarterly Reviewer of this Memoir on the 'Geology of Central France,' which was shortly after published.† That article was, I believe, the first essay of my distinguished friend Sir Charles Lyell, in the path of geological generalisation which he has since so successfully pursued. And I have sometimes ventured to think that during its composition he may have imbibed that philosophical

---

* 'Considerations on Volcanos, &c.,' 1825.
† Quarterly Review for May, 1827.

conviction as to the true method of inquiry into the past history of the globe's surface, namely, through a careful study of the processes actually in operation upon it, which is the leading principle of his deservedly popular works.*

---

* This was the pervading idea of both my early works, as will appear from the following passage in the Preface to the 'Considerations on Volcanos,' published in 1825.

"Geology has for its business a knowledge of the physical processes which are in continual or occasional operation within the limits of our planet, and the application of these laws to explain the appearances discovered in our geognostical researches, so as from these materials to deduce conclusions upon its past history.'

"The surface of the globe exposes to the eye of the geognost abundant evidence of a variety of changes which appear to have succeeded one another during an incalculable lapse of time. These changes are chiefly,—

"1. Variations of relative level between different constituent parts of the surface of the terraqueous globe.

"2. The destruction of former rocks and their reproduction under new forms.

"3. The production of new rocks upon the earth's surface.

"Geologists have hitherto usually had recourse, for the explanation of these changes, to the supposition of sundry violent and extraordinary catastrophes, cataclysms, or general revolutions.

"As the idea imparted by the terms cataclysm, catastrophe, or revolution, is extremely vague, and may comprehend anything you choose to imagine, it answers for the time as an explanation; that is, it stops further inquiry. But it has the disadvantage of stopping also the advance of the science by involving it in obscurity and confusion.

"If, however, instead of forming guesses as to what may have been the possible causes and nature of these changes, we pursue that which I conceive to be *the only legitimate path of geological inquiry*, and begin by examining the laws of nature which are actually in force, we cannot but perceive that numerous physical phenomena are going on at this moment on the surface of the globe, by which various changes are produced in its constitution and external character. These processes are principally: I. The atmospheric phenomena, including the laws of the circulation and residence of WATER on the exterior of the globe. II. The action of earthquakes and volcanos. And the changes *effected before our eyes*, by the operation of these causes on the constitution of the earth's crust, are chiefly,—

"1. Changes

PREFACE. ix

My purpose at all events was fulfilled. The Wernerian notion of the aqueous precipitation of "Trap" has since that date never held up its head. And I had good grounds for believing that the publication of the first edition of this Memoir, illustrated by an atlas of maps and drawings, which presented "oculis fidelibus" convincing evidence of the identic origin of ancient sheets of basalt crowning high mountain platforms, and contiguous lava-streams so recent as

---

" 1. Changes of level.

" 2. The destruction of some rocks and the reproduction of others from their materials.

" 3. The production of rocks *de novo* from the interior of the globe upon its surface.

" Changes these which, in their general character at least, bear so strong an analogy to those which appear to have occurred in the earlier ages of the world's history, that until the processes which give rise to them have been maturely studied under every shape, and then applied with strict impartiality to explain the appearances in question; and until, after a close investigation, and the most liberal allowance for all possible variations and *an unlimited series of ages*, they have been found wholly inadequate to the purpose, it would be unphilosophical to have recourse to any gratuitous and unexampled hypotheses for the solution of these analogous facts.

" The study of the processes by which these effects are at present produced on the surface of the globe, forms, therefore, a most important, but unfortunately most neglected, branch of geology."

I went on to say, that the work then produced was intended as a contribution to our knowledge of that division of the subject which relates to " the phenomena produced on the exterior of the globe by the development of its internal and subterranean activity," leaving to others, or to a future work, a corresponding inquiry into " the laws which determine the atmospheric influences, or the decomposition of rocks by air, light, electricity, or magnetism, and the conduct and mechanical effects of water on the surface of the globe, and on its solid parts."

Let me hint in passing, that this latter branch of inquiry has not yet, perhaps, been sufficiently pursued in detail, even by Sir Charles Lyell, or any other geologist.

still to appear cascading over the lips of craters of scoriæ, did assist in no small degree in putting an extinguisher on the then fading and flickering, but shortly before insolently triumphant, dogma. It had, likewise, I may hope, some share in directing attention to the vast influence exercised on the crust of the earth, not only by volcanic outbursts, but also by the erosive forces of rain and rivers, acting slowly and gradually, but through periods of immeasurable duration, upon the surface of supra-marine land.*

The edition of this Memoir printed in 1826 soon became exhausted. I was, however, unwilling to reprint it without previously revisiting the country of which the volume treats, and convincing myself of the accuracy of its descriptions; but it was only in the course of the past summer that I succeeded in accomplishing this.

Meantime Sir Charles Lyell and Sir Roderick Murchison, and many others, had followed me to Auvergne, and I had reason to believe were satisfied with the fidelity of my views and descriptions. The French geologists have also since that time paid more attention to this most interesting portion of their country than they had previously given to it. And I have been gratified to find many of the most trustworthy among them put forward views coinciding with

---

* See last note.

my own upon the several problems which there offer themselves for solution. I may instance M. Constant Prevost, whose ideas on the volcanic formations of that country, as expressed in a paper read before the Geological Society of France in 1843,* are identical with those which form the staple of my original volume.† So likewise are those of M. Bertrand de

---

* Bulletin XIV., p. 218–224.

† M. C. Prevost, as well as M. Pissis, and some other practical observers, have, with good reason, from the first, endeavoured to stem the current of opinion, which, among the Parisian geologists for some years so unfortunately set in favour of that most unphilosophical theory of "Elevation Craters," first suggested by M. de Buch, and afterwards warmly espoused by MM. Elie de Beaumont and Dufresnoy. Next to the exploded theory of Werner, I know of no fallacy that has so much impeded the march of true science, or been so obstinately persisted in. Such a notion could only find favour with geologists who had never witnessed the phenomena of volcanic eruptions on a large scale, and consequently had no conception of the normal mode in which their products are disposed. To one who has had that advantage, the theory does not appear to merit serious discussion. Pushed as it has been to extremity by the two last-named authorities, it in fact denies a volcano to be eruptive at all, that is to say, to be productive of any appreciable amount of lava or fragmentary matter; since such products, if they really accompanied in enormous quantities (as observers have been in the habit of believing from the evidence of their senses) the often repeated eruptions of active volcanos, must have accumulated about the orifice in the form of a mountain composed of those same multiplied beds of intermingled lavas and conglomerates, which these geologists persist in asserting not to have produced the mountain or its parasitic cones, which our eyes see them at work upon, but to have filled some supposed pre-existing hollow, afterwards raised by some sudden and anomalous process into the form of a mountain! It has never, of course, been denied by any sound writer on volcanic action, that a certain proportion of the frame of a volcanic mountain may have been partially elevated by those expansive throes or earthquake shocks which usually precede or accompany every eruption, and attest the production of fissures through its solid framework: which fissures being filled by the intumescent lava welling up from below, will no doubt when this is consolidated within them in the form of dykes, permanently raise and add in some degree to the bulk of the mountain, by what Sir C. Lyell

Doue, ably given in his description of the environs of Le Puy. MM. Le Coq and Bouillet have done me the honour to take many of my panoramic and other views as models for the illustrative engravings to their different works on the geology of Auvergne. Messrs. Rozet, Pissis, Ruelle, and others, many of whom were employed on the Government Geological Survey of the country, still in progress under the Ecole des Mines, have also printed or communicated to the Geological Society of France papers of interest on the tertiary and volcanic formations of Central France. Great light has likewise been thrown on the Palæontology of this district by the zealous researches and

---

aptly calls an "inward growth." Observations on the amount of matter composing the dykes, which are so frequent towards the centre of every volcanic mountain, would indicate perhaps a proportion of one-sixth or less of the substance of the central part of the cone as having been formed in this manner. And in some such proportion, elevation, in the sense implied by the elevation-crater theorists, may be admitted to have assisted in the production of the central summits. But at a distance from these few dykes are found, and there is no reason to doubt that the great bulk of such mountains owes in every case its formation to the heaping up of ejected matters, whether fragmentary or in the form of congealed lava-currents; a process which in fact may be witnessed in active volcanos, as the normal phenomena of every eruption. Any other view seems to me opposed not merely to the rules of philosophical analogy, but even to the evidence of the senses. The weight of authority unfortunately thrown into the scale in favour of this theory by geologists of such repute and influence as MM. de Beaumont and Dufresnoy has been the leading cause of the uncertain views and imperfect knowledge even now existing among French geologists on the great extinct volcanos of their own country. The clue to an understanding of a volcanic district must of course be a sound knowledge of the "modus operandi" of volcanic action; and this has been wanting in the modern school of Parisian geologists, with the exception of some few, who, like MM. Prevost and Pissis, above mentioned, have had the courage to oppose the influence exercised by two or three "great names."

able publications of MM. Croizet, Jobert, Bravard, Aymard, and Pomel.

The publication of the 'Explication de la Carte Géologique de la France,' by MM. Elie de Beaumont and Dufresnoy, has not yet reached those chapters in which the preface professes an intention to treat of the tertiary and volcanic formations of Central France. M. Le Coq is, I believe, engaged on a general work of this nature on Auvergne, to be illustrated by a geological map. But no portion of this work either has as yet been made public.

Under these circumstances, none of the publications, whether of French or English writers, which have yet appeared can be considered to afford that general view or detailed description of the very remarkable series of geological facts presented by this country which they undoubtedly merit, or which any visitor desirous of examining its phenomena would wish to have in his hand as a guide. I have therefore been led to suppose that a new edition of my Memoir, with such emendations and additions as time and the further observations, whether of myself or others, might suggest, would be acceptable at the present time. It will be seen that though I have found it desirable to recast some of the introductory chapters, the body of the work is still the same as was printed in 1826. Indeed on my late visit, I found no reason to alter the conclusions I had come to in 1821. Not possessing sufficient acquaintance with

Palæontology to venture on determining, of my own authority, the specific characters of the organic remains found in association with the rocks, whether tertiary or post-tertiary, of the district, I have given catalogues of its fauna from the works of MM. Pomel and Aymard in an Appendix, to which the text refers at the proper place. The illustrations have been recast and engraved on wood on a reduced scale, for the advantage of their being folded into the compass of an 8vo. volume, instead of necessitating, as in their original shape, an inconvenient accompanying folio atlas.

Of the maps, that of the Chain of Puys west of Clermont is nearly in the same state in which it appeared in the first edition. And I may add that it was entirely compiled from my own observations on the spot, on the basis of a sheet of Cassini's old Survey, as I had no access at the time to Desmarest's map, from which it has, I believe, been supposed that I took the details.

The other general map of Central France is copied from the 'Carte Géologique de France,' of MM. de Beaumont and Dufresnoy.

CONTENTS.

CHAPTER I.
Sketch of the Geology of Central France 1

CHAPTER II.
Tertiary Lacustrine Formations 6
    1. Limagne d'Auvergne. 2. Cantal. 3. Haute Loire. 4. Montbrison.

CHAPTER III.
Introductory Account of the Notices which have been hitherto published concerning the Volcanic Remains of the Interior of France 30

CHAPTER IV.
General Account of the Volcanic Formations occurring upon the elevated Granitic Platform of Central France   37

CHAPTER V.
First Volcanic Region. — Monts Dome and the Limagne d'Auvergne 40
    1. Chain of Puys. 2. Products of earlier Eruptions.

CHAPTER VI.
Region II.—The Mont Dore 114
    § 1. The Volcanic Mountain—Its general Outline—Conglomerates.
    § 2. Structure — Central Peaks — Trachytic Plateaux — Clinkstone and Basalt.
    § 3. Recent Eruptions.

CHAPTER VII.

Region III.—Cantal 145
  § 1. The Volcanic Mountain.   § 2. Canton d'Aubrac.

CHAPTER VIII.

Region IV.—Departments of the Haute Loire and Ardèche   154

  1. Mont Mezen: vast Currents of Clinkstone and Basalt—Breccias.
  2. More recent Eruptions—Subsequent Excavation of Valleys—Human Bones in Volcanic Tuff—Volcanos of the Haut Vivarais.

CHAPTER IX.

Concluding Remarks 197

  Age of Volcanic Rocks—Gradual Excavation of Valleys—Original limits of Lake-basins—Probable Changes of Level.

APPENDIX.

Catalogues of Organic Remains 217
Table of Heights 232
Explanation of the Maps and Engravings 234

Index 247

LIST OF ILLUSTRATIONS.

Plate.		
VI. Valley of Villar and Plateau of Prudelle	*To face page*	105
VIII. Clinkstone Rocks, Tuilière and Sanadoire, from the Puy Gros (Mont Dore)	,,	136
X. Montagne de Bonnevie (a cluster of Basaltic Columns), above the Town of Murat (Cantal)	,,	150
XII. Basaltic Plateaux of the Coiron (Ardêche), from the South	,,	163
XIV. Valley of Montpézat (Ardêche)	,,	187
XV. La Coupe d'Ayzac (Ardêche)	,,	194
XVII. Profiles and Sections, &c.	,,	205
XIII. Volcanic Cone and Basaltic Lava-current of Jaujac (Ardêche)	*Frontispiece.*	
I. Panoramic View from the Puy Girou, 6 miles South of Clermont (Puy de Dôme)	*At the end of the Volume.*	
II. General View of the Chain of Puys, from the West, above the Valley of the Sioule, near Pont Gibaud	,,	
III. Transversal View of the Monts Dôme, from the summit of the Puy Chopine	,,	
IV. Eastern View of Monts Dôme, from the Croix de Pirobot, between Volvic and Channat	,,	
V. The Southern Chain of Puys, from the Puy de la Rodde		
VII. Valley of the Dordogne and Mont Dore, from the Puy Gros on the North	,,	
IX. The Valley of Chambon and the Mont Dore from the East	,,	
XI. Panoramic Sketch of the Basin of Le Puy (Haute Loire) and of the Mont Mezen, taken from the Mont d'Ours	,,	
XVI. Sections of Granitic Plateau from East to West and from N.N.E. to S.S.W.	,,	
MAP of Central France	,,	
MAP of Puys of Auvergne	,,	

GEOLOGY AND EXTINCT VOLCANOS

OF

CENTRAL FRANCE.

CHAPTER I.

GRANITIC PLATEAU AND MARINE STRATIFIED FORMATIONS.

The parallel of 46·30, passing near the towns of Châteauroux and Châlons-sur-Saône, will be found to divide France into two nearly equal portions, of which the northern may be considered as a vast plain, whose waters flow gently towards the north and west through the Seine and the lower Loire. South of this line the surface continues to rise with a gradual slope, so as to form an inclined plane, which progressively acquires an elevation of more than 3000 feet above the sea in the Auvergne and Forèz, and a still greater in the Gevaudan and Vivarais, where it reaches the height of 5500 feet. Here it is abruptly cut down by the deep valley of the Rhone, which, running nearly due north and south, separates it from the ranges east of that river, in the departments Drôme, Isère, and Hautes Alpes. On the south-west also this high ground rapidly descends through broken and irregular embranchments to the basin of the Gironde. It may, in fact, be considered as a triangular platform, tilted up at its south-eastern angle, and declining gradually to the north-west. The principal mass of this elevated district is composed of primary crystalline rocks, chiefly granite, overlapped on all sides by secondary strata belonging principally to the Jurassic system, which, at its southern extremity, attain a considerable elevation in the chain of the Cevennes. Abstraction is in this description made of the volcanic products which rise, like enormous protuberances, from the higher parts of the elevated granitic platform. It is also deeply indented by the valleys of the upper Loire and Allier. These, on some points, acquire considerable width; the first in the basins of Montbrison and Roanne, the latter in the plain of the Limagne. Several detached basins of carboniferous sandstone occur within this district, seeming to have been deposited in hollows or estuaries of the original island of primary rocks. And there are vestiges of four geographically distinct deposits of as many freshwater lakes belonging to the tertiary period, occurring severally in Auvergne, the Forèz, the Cantal, and the Velay.

The granite of this district varies much in character, often within very narrow limits passing into gneiss, and sometimes, especially on its southern and western borders, into mica-schist, talcose-schist, or serpentine. The mica is sometimes replaced by pinite, either in amorphous nodules or crystallized in hexagonal prisms. It is here and there traversed by veins and dykes of fine-grained granite, of compact felspar, and of felspar porphyry. The felspar of the granitic rocks sometimes takes the form of large twin crystals, occasionally rose-coloured, like those of Baveno. The quartz often also presents beautiful crystallizations. The amethysts of Le Vernets, near Issoire, have long been known in commerce. In metals this primary district is not rich. Iron is very generally disseminated, but is only worked on a large scale at Alais (Gard), and at Rive de Gier, in the coal-basin of St. Etienne (Loire). Near Pont Gibaud argentiferous sulphuret of lead occurs, and has been wrought lately at a considerable expense, but, it is believed, with no great success. The same ore occurs near Villefort (Ardèche), and in the departments Aveyron and Lot, generally accompanied with manganese. The granite round Ardes (Puy de Dôme) and Massiac (Cantal), and the mica-schist of the Lozère, are rich in antimony. Copper is rare, but some veins in the Aveyron are supposed to have been anciently worked by the English when they were in possession of the country. Near Limoges and St. Yrieix the gneiss rock is decomposed into a kaolin of great purity, which has long supplied the china factories of Sèvres and Paris, and is even exported to the United States. Generally, the granite decomposes readily on the exposed surfaces, and presents therefore rounded outlines; while the gneiss, containing more quartz and mica, and having a schistose divisionary structure, exhibits peaked eminences and precipitous escarpments.

The mica-schist passes on some points into clay-slate, as near Alassac (Corrèze). With the only other exception of the very limited district of Tarare, between the Rhône and Loire, where a quartziferous sandstone, probably Devonian, and accompanied with anthracite, has been penetrated by a large outburst of red porphyry, the entire region of Central France contains, I believe, scarcely any sedimentary strata more ancient than the carboniferous; the Cambrian, Silurian, and Devonian series being generally absent. The coal-measures are sometimes associated with triassic strata, but much more usually with lias, and other members of the Jurassic system. In connection with these strata, both arenaceous and calcareous, coal is found in detached patches nearly all round the granitic platform, and, as already observed, on many points within its limits, especially on one straight line crossing it from north-north-east to south-south-west, from near Moulins to Mauriac. The direction of this line is remarkably coincident with the apparent axis of the granitic dome, and with the neighbouring volcanic range. The most important of these coalfields are at Autun (Saône et Loire); Decize (Nièvre); Villefranche and Bert (Allier); Brassac, in the basin of the Allier, near Lempde; St. Etienne and Rive de Gier (Loire); Lavoulte, Prades, and Joyeuse (Ardèche); Alais and Garges (Gard); Creuze and Bedarrieux (Hérault); Sansac, Layssac, and Aubin (Aveyron); Brives (Corrèze); Bourglastic and Bassignac, in the basin of the Dordogne; Bort and St. Eloy (Puy de Dôme).

An extensive series of limestone strata, belonging to the lias and oolite group, embraces (as has been before observed) the whole granitic platform like a frame; on its southern border especially, these calcareous rocks assume a remarkable development, forming a large proportion of the surface of the departments Aveyron, Lozère, Gard, and Ardèche. They constitute a vast elevated platform, sloping gradually from the primary range towards the south-west, and intersected by a few deep gorges, scarcely wider anywhere than the bed of the river that flows at the bottom of each. The stratification being nearly horizontal, though dipping to the south or south-west, this formation exhibits a series of flat-topped hills bounded by perpendicular cliffs 600 or 800 feet high. These plateaux are called "causses" in the provincial dialect, and they have a singularly dreary and desert aspect from the monotony of their form and their barren and rocky character. The valleys which separate them are rarely of considerable width; winding, narrow, and all but impassable cleft-like glens predominate, giving to the "Cevennes" that peculiarly intricate character which enabled its Protestant inhabitants in the beginning of the last century to offer so stubborn and gallant a resistance to the atrocious persecutions of Louis XIV.

The lias underlying the oolitic beds is often represented by blue schistose marls or sandstone, and occasionally by magnesian limestone, especially in the departments of Dordogne, Lot, and Aveyron.

CHAPTER II.

TERTIARY FORMATIONS.

Some three or four depressions in the surface of the elevated granitic tract described above appear to have been occupied, from an early period in the tertiary æra, by as many freshwater lakes. These have left the proofs of their former existence in sedimentary strata of clay, marl, limestone, and sandstone, frequently containing the remains of numerous species of freshwater mollusks, under such circumstances as to show that they were deposited for the most part in tranquillity along the shores and at the bottom of these basins, until they had accumulated to a thickness of several hundred feet. But the lakes have long since been drained—probably through the effect of some of the subterranean convulsions of which this country has evidently been the theatre; and these lacustrine strata have been cut through, worn down, and degraded by the usual denuding agencies, pluvial and fluviatile, until but small portions of them remain to tell the tale of their former enormous development. The original boundaries of these several lakes, though not to be laid down with perfect accuracy, may be more easily recognised than is usually the case with such formations, since their lowest deposits are seen almost everywhere resting against the granite ranges which, no doubt, formed their shores. The accompanying map marks with sufficient accuracy the general disposition and limits of these lacustrine formations.

I. Lacustrine Formation of the Limagne d'Auvergne.

The largest of these lakes covered the surface now occupied by the wide and fertile valley of the Allier, called la Limagne d'Auvergne, extending from Brioude in the south to some distance beyond Moulins on the north, having an average width of near twenty miles. It was bounded laterally by two parallel granitic ranges, that of the Forèz, which divides the waters of the Loire and Allier on the east, and that of the Monts Dômes, which separates the Allier from the Sioule on the west. An arm of the same lake also evidently reached some way up the valley of the Loire from the junction of that river and the Allier to the neighbourhood of Roanne. The lower levels of this great valley-plain are for the most part superficially covered with alluvium, composed chiefly of pebbles of granite, gneiss, trachyte, and basalt.* The substratum, wherever it shows itself, consists of nearly horizontal strata of sand, sandstone, calcareous marl, clay, or limestone, none of which observe a fixed and invariable order of superposition. Several hills also composed of the same strata rise from the plain, generally connected more or less with the granite ranges on either side. They were very justly described by M. Ramond, writing in 1815,† as "scattered relics of a series of beds which once covered the actual surface of the valley, and constituted an ancient plain much above the level of the present one."

Many of these fragments have evidently been protected from the destruction that has involved the greater part of the formation by a covering of basalt; others owe their preservation to a

---

* The compilers of the 'Carte Géologique de la France' class these alluvial beds into two divisions, one supposed to be of recent origin, the other referred to the Pliocene or upper Tertiary period, from its containing fewer pebbles of basalt, and generally occurring at higher levels, than the existing river-beds. But these distinctions do not seem to me as yet well made out.

† Mémoire sur le Nivellement des Monts Dômes.

similar capping of horizontal strata of a hard and durable limestone of a stalagmitic character to be described hereafter. These hills are seldom found in immediate union with the granite ranges, but are in general separated from them by shallow transverse valleys, so that the precise junction of the two formations is not often observable. But where this can be seen the lowest arenaceous beds are usually found to lean against the granite, sometimes at a considerable angle.

The principal divisions of the lacustrine series may be classed as—1st, grit and conglomerate, associated with red, blue, and white marls and sandstones; 2ndly, green and white foliated marls; 3rdly, limestone or travertin, often oolitic.

1. The sandstones and conglomerates forming the first or lowest of these divisions were considered by M. Brongniart, who gave them the name of "Arkose," to be of secondary and marine formation, and to be much anterior to the lacustrine strata with which they are associated; and several of the French geologists have followed him in this view. As they rarely contain any definite organic remains, either alternative seems at first tenable. But on a closer examination there is no difficulty in recognising the occasional alternation of these sandstone-beds with the calcareous strata containing freshwater shells,* and I have therefore no doubt of their belonging to the lacustrine series. They usually are seen resting directly on the granite edges of the basin, from the detritus of which they are evidently derived. Some beds consist of a conglomerate of worn pebbles and fragments of granite, gneiss, mica-schist, porphyry, the rocks of the adjoining elevated district, but without the admixture of basaltic or any other volcanic rocks. These arenaceous strata are not continuous round the margin of the lake basin, being rather disposed here and there, like the independent deltas

---

* For example, in the hills called les Côtes and Chanturgue near Clermont.

which grow at the mouths of torrents, along the borders of existing lakes.*

Other beds consist of a quartzose grit formed of separate crystals of quartz, mica, and felspar, evidently formed *in situ* from the disintegrated materials of the granite on which they rest, and from which they are scarcely to be distinguished. The cement is generally siliceous, and the resulting rock is sometimes hard enough to be used for millstones. At other times the cementing matter is calcareous, and the stone more brittle. The calcareous matter sometimes is aggregated into nodular concretions, passing into solid beds of limestone resembling the Italian travertin or the deposits of mineral springs. It occasionally contains crystallized sulphate of barytes in veins, chalcedony, and bitumen. The most largely developed of these arenaceous beds are, however, those of red, blue, and yellow marls and sandstones, which present an aspect absolutely identical with the secondary *new red* sandstone and marl of England. Some of these strata are very friable, others sufficiently compact to be quarried for building-stone. They are, like the conglomerates above mentioned, with which they occasionally alternate, evidently derived for the most part from the degradation of the adjoining granite or gneiss, which is in fact seen to decompose into an alluvium very similar to these tertiary sands and marls. The calcareous element, where it occurs, was, no doubt, added from the interior of the primary crystalline rocks, whence even now issue many springs depositing large quantities of carbonate of lime, and which must have been the source of the far greater bulk of that material composing the marly strata generally superimposed to the sandstones.

2. Green and white foliated marls. Sir Charles Lyell observes

---

* Lyell, Man., p. 698.

that the same primary rocks which, by the partial destruction of their harder parts, gave rise to the quartzose grits and conglomerates before mentioned, would, by the reduction of the same materials into powder or fine mud, and the decomposition of their felspar, mica, and hornblende, produce aluminous clay, and, by admixture with carbonate of lime from the springs, calcareous marl. This fine sediment would naturally be carried out to a greater distance from the shore than the coarser materials which are usually found in the immediate neighbourhood of the granitic borders. These chalky marls certainly attain a thickness in some spots of 600 or 700 feet. They are for the most part either light yellowish green or white, and have very much the aspect of chalk, with, like it, a semiconchoidal fracture when the strata are sufficiently thick. Usually, however, they are thinly foliated—a character which arises from the innumerable thin shells, or carapace valves, of that minute animal called cypris, which is known to moult its integuments periodically, differing in this from the conchiferous mollusks. On other points flattened stems of charæ, or myriads of small paludinæ or other freshwater shells, may be observed by the microscope to occasion this foliation, which is carried to such a degree that twenty or thirty laminæ may often be counted in the thickness of an inch.

3. Interstratified with the marls we find thick beds of an oolitic limestone resembling our Bath stone in colour and structure, and, like it, acquiring greater hardness on exposure to the air. At Gannat and elsewhere this rock contains land-shells and bones of quadrupeds and birds. At Chadrat the oolitic grains are so large as to deserve the name of pisolites, the small spheroids combining both the radiated and concentric structure.

But the most remarkable form assumed by this freshwater limestone is that called "Indusial," from the cases, or indusiæ, of caddis-worms (the larvæ of Phryganeæ), great heaps of which have been incrusted as they lay by carbonate of lime, and formed into a hard travertin or stalagmitic limestone. This rock is seen sometimes to form ranges of concretionary nodules, at others continuous beds, one over another, with layers of the foliated marls interposed.

It is well known that certain varieties of the Phryganea (or caddis-fly) are in the habit, when in their caterpillar state, of clothing their bodies with a cylindrical case composed entirely of minute river-shells of some single species—helices, mytili, planorbes, or other—united by glutinous filaments, and disposed in some sort of order around. These habitations are quitted when the insect's metamorphosis is completed; and on the banks of rivers or marshes frequented by them, heaps of such empty cases may be observed. If we suppose them in this state to be exposed to be incrusted by calcareous matter from the depositions of some neighbouring spring, they will assume precisely the appearance of the remarkable rock which we find in Auvergne, composing repeated strata of considerable bulk, alternating through a thickness of several hundred feet with the more ordinary marls. The surfaces of these beds are usually mammillated or botryoidal, and the calcareous matter enveloping the Indusiæ is arranged concentrically in the manner of a stalagmite. Where the bed is thinnest, the continuity is often interrupted or prolonged in separate nodular concretions of the same kind imbedded in loose sand. The minute shells surrounding the larva-cases are usually the Bulimus atomus of Brongniart, or a small Paludina. More than a hundred of these shells may be counted round a single tube, and ten or twelve tubes may be found packed together irregularly in a single cubic inch of the rock. When it is added that *repeated* strata of this kind eight or ten feet thick appear to have covered very many square miles—originally, perhaps, indeed the whole plain of the Limagne, measuring forty miles by twenty—some idea may be formed of the countless myriads of minute animals belonging to one or two species only of mollusks which must have formerly lived and died on the bottom or shores of this extensive lake! The places where I have observed these characteristic beds best developed are on the hills of Gergovia above Romagnat, at the Puys Girou, de Jussat, de la Serre, de Monton, de Dallet, at Mont Chagny, Mont Jughat, and Les Côtes near Clermont; at Davayat near Riom; at Aigueperse, Gannat, Mayet d'Ecole, St. Gerard-le-Puy, between Jaligny and la Palisse, at Pont Barraud, &c. It is unnecessary to suppose that the Phryganeæ or the mollusks all lived precisely on the spot where their remains are found. They may have multiplied in the shallows near the margin of the lake, or in the streams by which it was fed, and the cases have been drifted into the deep water, perhaps (as is suggested by Sir Charles Lyell) borne upon the masses of littoral reeds to which they were attached, when these were torn away from the banks by storms, and floated out by the winds and currents to great distances.* Sometimes, instead of the Indusiæ, the calcareous incrustation appears to have enveloped the reeds themselves, grasses or mosses growing on the bank. At others the beds lose their stalagmitic character, and appear as a compact earthy limestone of an ochreous yellow colour, and contain large shells of the genera Helix, Planorbis, or Lymneus. The testaceous matter is often replaced by calcareous spar, occasionally by bitumen, which appears too in veins or crevices of this rock. Some are traversed by veins of flint or semi-opal. Indeed the whole rock becomes in some parts highly siliceous, breaking into sharp splinters with a conchoidal fracture. And

---

* Manual, p. 202, ed. 1856.

some resemble in mineral character the secondary lithographic limestone of Châteauroux. The calcareous marls often contain much gypsum, selenite filling the fissures of the rock and of the associated marls; and that so abundantly as to be extracted for commercial purposes — as at the Puys de Coran, de Millefleur, and the Butte de Montpensier.

In general it is in the more solid strata just described, rather than in the large intervening deposits of white, soft, foliated marls, that are found the remains not only of mollusks, but also of the vertebrated animals, of which a very copious catalogue has been published by M. Pomel.* Carnivora, Insectivora, Rodentes, Ungulata in numbers; and among them the anthracotherium, rhinoceros, dinotherium, cœnotherium, and palæocherus; Ruminants, crocodiles, tortoises, lacertæ, birds, serpents, fish, and batrachians are also frequent. All these, says M. Pomel, seem properly to belong to those great deposits of concretionary indusial limestone which form so important an element in the geognosy of the whole Limagne.† This is in fact what would be expected from the littoral origin we have already ascribed to the contents of these beds. M. Pomel declares the result of the examination of his catalogue to be that all the ossiferous deposits of the Limagne lake belong to the same geological and zoological period (the Lower Miocene), and cannot be divided into periods by reason of their palæontological characters, since the same species are found in the oldest as in the most recent beds of this formation. Whatever differences occur are such only as are found in the existing fauna, and may be referred to differences in geographical or climatic situation, in the habits or the greater or less extension of species, to accidents of deposition or fossilization. With

---

* This catalogue will be found in the Appendix.

† Pomel, Catalogue des Vertèbres Fossiles, découverts dans le Bassin de la Loire supérieure et de l'Allier. Paris, 1854, p. 151.

respect to the question of correspondence in point of age of the lacustrine deposits of the Limagne with other well-known tertiary formations, M. Pomel exhibits more doubt, but on the whole concludes from palæontological evidence that they may be considered as parallel to the fossiliferous beds of Mayence, more recent than the gypsum of the Paris basin, but older than the Faluns of Touraine—a classification which corresponds very closely with the opinion of Sir Charles Lyell, who, after hesitating for some time as to whether these Auvergne deposits belonged to the upper members of his Eocene formation, or to the lower ones of the Miocene, has, I have reason to believe, made up his mind to consider them as of the age of the latter division of the tertiary series.*

These remarks refer, of course, to the lacustrine strata alone of the Limagne, not to certain later ossiferous alluvial deposits, such as occur at the noted localities of Mont Perrier and Pardines, which belong to a period when the volcanos of this country had been long in activity, the lake certainly drained, its strata largely denuded, and deep and wide valleys channelled out through them. To these I shall revert in a later page.

No traces of rocks of a volcanic character occur in the conglomerates or sandstones which constitute the lower terms of this freshwater formation. But higher in the series a mixture is here and there to be observed of the products of neighbouring volcanic eruptions with the calcareous beds. In occasional instances fragments of basaltic lava, crystals of augite, scoriæ, and volcanic ashes are scattered through some of the undisturbed limestone-beds, and assume frequently a remarkable disposition, the heavier and larger fragments occupying the lower part of each stratum, the lighter and smaller stopping in

---

* See the Supplement to his fifth edition of the Manual (Murray, 1857).

the upper parts—suggesting the obvious idea that these fragments, after ejection from a volcanic vent, had fallen through the air into the lake at the time the stratum in which they occur was in the condition of very soft calcareous mud. Geologists will probably see in this fact some analogy to the arrangement of the layers of flint nodules in common chalk strata.* I have observed strata of this character on the mountain of Gergovia and several other points; but the most remarkable example may be seen on the banks of the Allier immediately above Pont

1. Cliff on the Banks of the Allier, near Pont du Château, at the base of the Puy de Dallet.

du Château. The river here flows at the foot of a line of cliff, which it is continually undermining, and thus exposes a regular vertical section rather more than 100 feet in height. (See the woodcut above.)

The upper surface of the cliff is horizontal, and formed to the depth of 8 or 10 feet of a bed of boulders exactly the same as those which the river still rolls along its bed below. The strata beneath are not completely horizontal, but have a slight curve rising upwards at each extremity of the exposed cliff, as if they had been deposited in a hollow, or perhaps subjected to some amount of disturbance since their deposition.

---

* In which case likewise "it seems as if there had been time for each successive accumulation of calcareous mud to become partially consolidated, and for a re-arrangement of its particles to take place (*the heavier silex sinking to the bottom*), before the next stratum was superimposed."—*Lyell's Manual*, 5th ed., p. 244.

1. The lowest visible beds consist of a compact brownish grey limestone containing much fragmentary volcanic matter, generally arranged as above described in each stratum according to their respective gravities, the largest fragments at the bottom, the rest diminishing in size in proportion to their distance from thence. These beds dip rapidly below the level of the river, and are covered by—

2. A bed about 4 feet thick of a coarse-grained limestone of an ochreous yellow colour and earthy fracture, without any visible admixture of volcanic matter. It contains the casts of numerous shells, Planorbes, Helices, Lymnei, the shells being often replaced by bitumen, which exudes likewise from fissures in the rock. This is a very common variety of the hard limestone beds frequent throughout the lacustrine series.

3. Above this is a considerable thickness of the soft white foliated marly limestone, characteristic of the formation, having a plain fracture, sometimes slightly conchoidal, strongly adhering to the tongue, with an earthy smell, its exposed surfaces occasionally presenting efflorescent carbonate of soda. It has frequently a tendency to desquamate in globular masses.

4. A single bed, 10 inches thick, of a very compact, hard, and fine-grained limestone, with a conchoidal fracture, of a deep bluish-black, probably coloured by volcanic ashes.

5. Numerous calcareo-volcanic strata, similar to No. 1; some of them being ribboned with different colours, yellow, brown, or bluish-black, determined, it seems, by the greater or less proportion of volcanic ashes contained in their several zones. Where this proportion is considerable the rock generally has the more compact character and conchoidal fracture; and by still further increase of the volcanic element it passes into a kind of peperino or basaltic breccia, which often affects a globular concretionary structure.

The strata interspersed with volcanic matter are perfectly parallel to those that are free from it, and present all the characters that we should expect in sediment slowly and tranquilly deposited in a body of water into which repeated showers of volcanic ashes and fragments were occasionally ejected from some neighbouring volcano in active eruption. Immediately behind this cliff rises the Puy de Dallet (6 in woodcut), an isolated hill about 900 feet high above the river, composed entirely of repeated strata of the freshwater limestone and marls, with the exception of a heavy capping of basalt. And in the ravines with which it is scored, sections are afforded sufficient to make it abundantly plain that the beds already described in the cliff alongside the Allier pass under the entire hill.

Some of the strata composing the Puy de Dallet are very siliceous, some oolitic, and the upper beds are of the concretionary indusial variety. The basaltic platform rests immediately on a thick bed which contains much volcanic matter, and is in fact a basaltic breccia, mixed with calcareous particles. It is compact, and seems at a distance irregularly columnar, as is the superimposed basalt.

Here we have unquestionable proof that volcanic eruptions of basaltic lava and scoriæ occurred within or on the banks of the freshwater lake of the Limagne long before it had ceased to deposit sediment; since a thickness of several hundred feet of its sedimentary beds are found overlying some which contain numerous fragments of volcanic character. In the mountain of Gergovia a similar alternation is to be seen of beds of limestone and marls with others containing numerous fragmentary volcanic matters, often in such abundance as to compose far the greater part of the rock and give it the character of a peperino. This hill also is capped by an enormous platform of basalt: and another thick bed of the same rock crops out from the side of the hill about a hundred feet lower, the interval being almost wholly composed of horizontal strata of calcareous peperino. It has been questioned whether this latter bed of basalt is a true lava-current like the sheet which caps the hill above it, or is not rather an intruded dyke that has forced itself between the strata; a distinction of no great importance. It probably partakes of both characters; certainly it is associated with several true dykes which penetrate the calcareous strata beneath it. There can be no doubt, however, of its having been formed before many of the calcareo-volcanic beds which cover it, and which exhibit all the marks of sedimentary deposits accumulated under circumstances of great disturbance from neighbouring and contemporary eruptions.*

On other spots of the lake-basin masses of calcareous peperino occur in which stratification entirely disappears. All is confusion, the rock passing by frequent and gradual transitions from a limestone impregnated but slightly with volcanic particles to a calcareous peperino, *i. e.* conglomerate composed of fragmentary basalt and scoriæ, cemented by calcareous spar, and finally to a compact basalt.

Such examples are found in the Puys de Crouel and de la Poix (remarkable for its abundant bitumen), at Verthaison, Cour-cour, Pont du Château, and the Puy Marmant. In these cases the rock appears to be the result of local volcanic eruptions through the still soft calcareous mud which then formed the bottom of the lake. It is reasonable to suppose that there would have been under these circumstances the most intimate admixture of the calcareous matter with the erupted igneous rock, such as we find here.

At the first of these localities, the hill of Verthaison, a great

---

* See the description of Gergovia, p. 107 *infra*.

quantity of fine specimens of radiated arragonite occurs, in veins, some of which are a foot in thickness. They intersect the rock in such numbers as to give it a reticulated appearance. A range of hill consisting solely of the same calcareous peperino is continued for about two miles towards the south.

Above the village of Chauriat it presents a very compact and hard stone, composed of small unequal fragments of augitic basalt firmly united by a fine cement of calc spar. It is difficult, while examining this rock, to believe that the basaltic parts are in reality mere fragments, and the impression is strongly excited that its peculiar structure is owing to a separation of dissimilar substances during the consolidation of a lava which by some mechanical cause had been universally penetrated with calcareous matter.*

La Montagne de Cour-cour, an insulated hill of considerable size rising alone from the plain near Beauregard, is of the same nature. The peperino here is traversed by irregular venous masses of dark-grey basalt. Arragonite is not so abundant as at Verthaison.

---

* The calcareous peperino of the Vicentin (Montecchio Maggiore) exhibits some very similar mixtures of basalt and calcareous spar, which it is difficult to refer decidedly either to the conglomerate or the solid lava-rock.

In giving the name of *peperino* to a volcanic conglomerate consisting of fragments of basalt and scoriæ, without pumice or any trachytic matter, united either by simple adhesion or a calcareous or argillaceous cement, I follow the Italian geologists, who have continued this trivial term to a similar rock, which also, like that under consideration, occasionally contains fragments of limestone and primitive rocks, bituminized wood, &c. &c.—*Vide* Brocchi, Catalogo ragionato di Rocce, pp. 45, 47.

There exists the strongest analogy between the calcareous peperino of the Limagne and that of the Vicentin, the latter being without doubt the result of volcanic eruptions breaking forth from the bottom of the *sea*, in which vast masses of calcareous matter (of the Pliocene tertiary formation) lay in a pulpy unconsolidated state; the former of eruptions through a similar mass, the deposit of a *freshwater* lake. The great variety of rare and beautiful crystallizations to which this violent mixture of calcareous matter with incandescent lava has given rise in both localities, is remarkable. Mesotype, stilbite, arragonite, chalcedony, and numerous forms of calcareous spar, abound in the drusy and vesicular cavities and veins of both these conglomerates.

The Puys de la Piquette and de Marmont, which rise on the other side of the Allier above Vayre, the first relay on the posting-road from Clermont to Le Puy, are similar examples of amorphous masses of calcareous peperino traversed by vertical dykes of basalt. The latter hill is connected by the base with another of a rather larger size, on which stands the town of Monton, but differs from it in substance. Its principal component rock is like those already described, a peperino of basaltic fragments, small scoriæ, and fine volcanic detritus mingled with fragments of limestone of various sizes, and united by a calcareous cement. On the western side of the hill fronting the Puy de Monton the limestone fragments predominate; the opposite or eastern face consists of an amorphous mass of basalt, hard, solid, and of a dark bluish grey colour, which has here without doubt burst from below through the calcareous sedimentary beds. Parts of this rock are amygdaloidal, being thickly larded with globules of compact radiated mesotype; this mineral having evidently occupied the vesicles formerly existing in the stone. Where the cavity is of a considerable size, it is not filled throughout with the zeolite, but a nest or hollow pouch appears, lined with the most brilliant crystals. These are too well known in all mineralogical collections to require a description here. That variety is most common which consists of radiated groups of large quadrilateral prisms terminated by an obtuse pyramid whose faces correspond with the sides of the prism. Numerous capillary crystals resembling the finest spun glass are sometimes joined with these, but more frequently occupy separate cavities. The same crystallizations are found in cavities of the peperino as well as of the basalt; others present different beautiful varieties of calcareous spar.

The peperino of Pont du Château is equally rich in rare and splendid specimens, which are well known to all collectors of rare minerals.

The rock upon which stands part of the town is traversed by fissures from which a prodigious abundance of viscid bitumen exudes spontaneously. The cheeks of these clefts are frequently lined with mammillæ of chalcedony, and from these occasionally spring the most graceful groups of small rock-crystals diverging from a central point, and spread out with the regularity of the petals of a marigold. The union within a single specimen of the smallest dimensions, of volcanic scoriæ, limestone, chalcedony, rock-crystal, and bitumen, is singular.

Some of the upper parts of this rock present a spheroidal structure, similar to what often occurs in basalt. The spheroids are from two feet to a few inches in diameter. They exfoliate by decomposition in concentric coats like those of an onion. This configuration appears remarkable in a conglomerate, but is not unusual in the "wackes" (as they have been generally called) which accompany the older basalts. Veins of compact fine-grained and siliceous limestone, free from visible basaltic particles, though of a light-brown colour, pierce through this rock occasionally from below, and mix intimately with it at the line of contact.

The Puys de la Poix and de Crouel, and the low hill on which Clermont stands,* consist of a similar species of peperino, bearing equally the appearance of a violent and intimate union of volcanic fragmentary matter with limestone while yet in a soft state. They enclose ill-defined masses and veins of pure limestone, some bituminized wood, chalcedony, and bitumen. The latter substance formerly exuded in great abundance from the Puy de la Poix, but the source at present seems exhausted.

At the northern base of the hill upon which Clermont is built rises a spring, the water of which is impregnated by means of

---

* Nos. 90, 89, and 88, in our map of the Monts Dômes.

its carbonic acid with so large a proportion of carbonate of lime, which it deposits on issuing into the air, that its incrustations have formed an elevated natural aqueduct 240 feet in length, and terminating in an arch thrown across the stream it originally

2. Natural Bridge of Travertin, formed by an incrusting Spring at Clermont.

flowed into, 16 feet high and 12 wide. Near it are the rudiments of a similar arch, the construction of which is still going on, and aptly illustrates the formation of the other. Like many other similar incrusting springs in different countries, it has been turned into a source of emolument by the proprietor, who breaks the fall of the water in such a manner that its stony particles may be deposited on various natural objects exposed to its spray. At the time of my visit the stuffed skins of a horse and a cow were undergoing this petrifying process, as well as the usual proportion of birds, fruit, flowers, medals, cameos, &c.

There are several other springs in Auvergne possessed in a high degree of the same qualities. One at Chalucet, near Pontgibaud; another, called La Fontaine de Rambon, on the banks of the Crouze, near St. Floret. On both sides of this river, and for a considerable distance down the gorge it flows in, are seen colossal fragments of calcareous travertin, which by their position prove this mineral spring to have once erected a bridge similar to that of St. Alyre at Clermont, but exceeding it prodigiously in dimensions, and probably choking up the whole valley, since the source itself is elevated more than 100 feet above the river.

It is worthy of remark that the three springs mentioned above, whose deposits are, except in a greater or less proportion of iron, exactly alike, rise from rocks of different kinds: the first, from a calcareous peperino; the second, from the foot of a regular volcanic cone, at least twenty miles from any calcareous rock; the third, from granite. It is apparent from this that all have their origin in or below the granitic rocks which form the basis of the whole territory, and which include or cover the volcanic focus whence in reality these mineral springs in all probability ultimately derive. The same observation applies to the many *thermal* sources which occur on various points of the platform, springing indifferently from primitive or volcanic rocks; as at Mont Dore les Bains, La Bourboule, St. Nectaire, Châtelguyon, Gimeaux, Néris, Vichy, Vic en Carladèz, Chaudesaigues, &c. Some of these deposit a travertin having much silex as well as carbonate of lime in its composition, and arragonite is occasionally found to have crystallized in its fissures. There appears reason to believe that the quantity of mineral matter brought to the surface by such springs was much greater in earlier times, and is still annually diminishing.

II. Basin of the Cantal.

A freshwater formation, probably of the same age and very similar to that of the Limagne, occurs in the department of the Cantal, particularly in the neighbourhood of the chief town, Aurillac. Its principal distinction consists in the far greater abundance of silex associated with its calcareous marls and limestone. As in the Limagne formation, the lower beds are arenaceous, and derived apparently from the detritus of the rocks of gneiss and mica-schist within which the freshwater basin lay, and upon which these beds rest. The upper series consists of calcareous and siliceous marls, containing subordinate beds of silex, gypsum, and limestone. The bands of silex often very much resemble in position and aspect the flints of our English chalk, being sometimes continuous, sometimes in layers of concretionary nodules, assuming much the same forms as our flints, and, like them, having a white surface. Their substance is, however, generally more vitreous, like resinite, and often approaches to opal or agate. Its colours are usually yellowish-brown or greyish-blue, and it is sometimes ribboned delicately with stripes of these colours. The marly limestone, when pure from flint, is white or yellowish white, of an earthy and coarse fracture, full of shapeless tubular cavities and small filamentous perforations, such as would be left by reeds, grasses, or other weeds enveloped as they grew in a calcareous sediment or incrustation. It contains numerous shells of the genera Potamides, Helix, Limneus, Bulimus, Planorbis, &c.; with minute cypris and the seed-vessels of charæ. Sometimes the interior of the shell is lined with chalcedony, while the shell itself is of carbonate of lime; at others, the shell is of flint; the interior, limestone. The foliated marls are as thinly laminated as paper, owing probably to the myriads of small shells, or flattened stems of charæ, preserved in them. Several hills are seen in the neighbourhood of Aurillac composed of such strata two or three hundred feet in thickness. They are generally covered by massive beds of volcanic breccia, trachyte, and basalt. In some places, as between Aurillac and Polminhac, there appear to be a confusion, and occasional alternations, of the volcanic and calcareous beds, similar to what has been described in the Limagne formation.

The original limits of the lake-basin of the Cantal can with difficulty be ascertained, owing to the colossal proportions of the volcanic mountain in the vicinity, by whose products its deposits have been overwhelmed. They are found, however, beneath the volcanic beds wherever any torrent discloses the inferior strata, within a space limited by lines passing through Jussac, Vic en Carladèz, Mur de Barrèz, and Panet. But as they reappear on the north-east side of the central heights of the Cantal, near Murat, bearing the same characters as in the valleys of the Cère, the Goule, and the Jourdanne, it seems probable that the lake was continuous through the intervening area.

III. Basin of the Haute Loire.

The freshwater formation of the basin of the Upper Loire which surrounds Le Puy differs but slightly from the two already described, and, like these, has been covered in part by prodigious and repeated extravasations of volcanic matter, which have loaded its generally-horizontal strata with massive coverings of basalt and basaltic breccia three or four hundred feet in thickness. It is, therefore, in general only by tracing up the deep waterworn ravines which furrow these superimposed rocks that the extent of the lacustrine deposits can be observed: the main valley of the Loire, however, and some of its tributaries, offer an ampler view of the beds through which they have been excavated.

The limits of the original basin show themselves on the west, in the base of the granitic chain which separates the waters of the Loire and the Allier. The rise of the granite platform towards the Haut Vivarais bounded it on the south; and some irregular embranchments from the heights of St. Bonnet and La Chaise Dieu on the east and north. One of these granitic spurs indeed stretches across the lacustrine formation, and separates it into two, cutting off the small basin of Emblavès from the upper and larger one, in which lies the town of Le Puy. The river Loire passes now from the latter basin into that of Emblavès through the narrow, deep, and sinuous gorge of La Voute, and finally issues again from this by means of the similar defile of Chamalières. Both of these outlets appear to owe their origin to some of the most recent changes to which this singular district has been subjected. Certainly neither could have been in existence at the time that the massive sheets of basalt and clinkstone which severally cap the cliff-ranges above them were in a state of igneous liquefaction.

The lower series of lacustrine beds consist, as in the formations already described, of sandstone, blue, green, and red variegated sandy marls, and clays. The sandstone is an excellent building-stone; the clays are used for pottery.* The upper beds are chiefly of marly limestone, often highly siliceous, and enclosing

---

* Doubts have been expressed with respect to these sandstones, &c., as well as the analogous beds which underlie the unquestionably freshwater and tertiary marls and limestone of the Limagne, whether they do not belong to the secondary formation of new red or variegated sandstone. The remains of plants which they occasionally contain indicate a marshy soil and atmosphere, consisting of large reeds, casts of *cyclopteris* and *pecopteris*, with some small seeds and fruits which appear referable to dicotyledonous plants. No shells have yet been found. In the corresponding sands of Auvergne a species of Cyrena is met with, and M. l'Abbé Croizet is, indeed, said to have found a bivalve apparently of marine origin; but the evidence of this isolated fact seems obscure, and not to be depended on, unless confirmed by future discoveries of the same character. See Appendix for the Fauna of the Tertiary strata.

layers of flint passing into semi-opal, especially in the vicinity of volcanic rocks. At St. Pierre Eynac these siliceous strata evidently pass under the powerful clinkstone mass of Montplaux. Towards the middle of the basin the clayey marls alternate with beds of gypsum, several of which are sufficiently rich to be worked for agricultural and other uses. The shells contained in these beds are of lacustrine or marshy species; bones also occur in them, and the remains of fish, crustacea, birds, and their eggs.* Above the marls with gypsum is usually a considerable thickness of calcareous and foliated marly strata, alternating with greyish limestone of the consistence of chalk, having tubular cavities attesting its palustrine origin, and numerous casts of planorbes, limnei, cyclostomæ, bulimi, cyprides, &c. Bones and teeth of animals, both terrestrial and aquatic, are also found in them abundantly, for a catalogue of which I must refer to M. Pomel.† In one site alone, the hill of Ronzon, and nearly in one bed, so large a number and variety of organic remains occur as almost to furnish a complete Fauna of the district at the period of its deposition, which MM. Pomel, Aymard, and Lyell unite in referring to the Lower Miocene.

There is no clear evidence of the outburst of any neighbouring volcanos during the tertiary period in which these sedimentary beds were deposited in the freshwater lake of the Haute Loire. No alternation of volcanic matters with the sedimentary strata, such as those already described in the Limagne, have been detected within this basin. It is possible, therefore, that the earliest development of local volcanos occasioned such a disturbance as to cause the drainage of the lake; on the other hand, several of the basaltic breccias, which we shall have occa-

---

* M. Aymard mentions the following mammifers, as found by him in these beds:—*Palæotherium primævum, Pal. sub-gracile, Monacrum velaunum.*

† See Appendix.

sion to describe among the volcanic rocks of this district, bear a near resemblance to the peperino of the Auvergne lake-basin, and were perhaps, like that rock, the product of eruptions from within its area before the waters were wholly drawn off.

IV. Basin of Montbrison.

I have not myself visited this locality, nor am I acquainted with any detailed description of it. It occupies a valley-plain, about 20 miles long by 10 in width, encased between the granite and gneiss ranges of the Forèz and the Lyonnais on the south, west, and east, and the porphyry rocks and Devonian strata of the Tarare, through which the Loire finds an outlet to the north. The tertiary strata of this basin have so close a resemblance in character and position with those of the lower valley of the same river about Roanne, as to lead M. Raulin * to presume that the two basins were originally connected by a channel permitting their waters to maintain the same level. The same red and yellow sands, sandstones, clays, and green and white foliated marls are found at Marcilly, Boën, Sury le Comtal, and generally throughout the plain. Several eruptions of volcanic matter appear also to have taken place within this basin, and from the granitic heights to the west. But for the reason given above I am not aware of the precise circumstances under which the volcanic rocks present themselves. M. le Coq informs me that several basaltic dykes appear near the junction of the porphyry and granite; while others penetrate through beds of rolled pebbles, probably belonging to the lower term of the lacustrine series. A further examination of this basin seems very desirable.

V. Tripoli Basin of Menat.

At Menat, on the road from Riom to Montaigu, occurs a sin-

---

* Bulletin XIV. p. 584.

gular depression in the gneiss and mica-schist which here take the place of the granite of the primary or crystalline platform. It is nearly circular and about a mile in diameter, and discharges its waters into the Sioule through a narrow gulley worn in the schists not more than 12 feet wide and as many deep. Before this passage was effected they must have formed a lake over the whole valley, the surface of which is almost perfectly level.

The excavations that have been made in the sedimentary beds beneath this surface show them to be composed to a considerable but unascertained depth of bituminous shale, or desiccated flaky clay of a muddy black colour, evidently the fine detritus of the micaceous and talcose rocks which enclose the basin impregnated with bituminous matter, and often containing vegetable remains in such abundance as to become a true lignite. It envelops many nodules of iron pyrites, globular or lenticular, sometimes assuming the flattened moulds of fish, chiefly a cyprinus, very like the Cyp. papyraceus of the lignite of the Siebengebirge. The thin folia of the shale and lignite exhibit on their surface innumerable impressions of leaves resembling those of the chesnut, sycamore, willow, lime, and aspen, which still grow in the neighbourhood, with others which certainly do not belong to European species, and resemble those of the liquid amber, *Styraciflua* and *Gossypium arboreum*.* Some flattened fruit, or seed-vessels, are also found resembling those of the hornbeam. These lignites appear to have undergone spontaneous combustion on some points (probably where pyrites abounded), and the shale has been thus converted into reddish tripoli, which is largely quarried for commercial use. It seems probable that the formation of this simple alluvial deposit is of no very distant date.

---

* Bouillet, 'Vues et Coupes du Puy de Dôme.'

CHAPTER III.

INTRODUCTORY ACCOUNT OF THE NOTICES WHICH HAVE BEEN HITHERTO PUBLISHED CONCERNING THE VOLCANIC REMAINS OF THE INTERIOR OF FRANCE.

To those who now travel over the mountains of central France, and see on all sides marks of volcanic agency exhibited in the most decided manner, numerous hills formed entirely of loose cinders, red, porous, and scorified as those just thrown from a furnace, and surrounded by plains of black and rugged lava, on which the lichen almost refuses to vegetate, it appears scarcely credible that, previous to the middle of the last century, no one had thought of attributing these marks of desolation to the only power in nature capable of producing them. This apparent blindness is however very natural, and not without example. The inhabitants of Herculaneum and Pompeia built their houses with the lavas of Vesuvius, ploughed up its scoriæ and ashes, and gathered their chesnuts from its crater, without dreaming of their neighbourhood to a volcano which was to give the first notice of its existence by burying them under the products of its eruptions. The Catanians regarded as fables all relations of the former activity of Ætna, when, in 1669, half their town was overwhelmed by one of its currents of lava.

In the year 1751 two members of the Academy of Paris, Guettard and Malesherbes, on their return from Italy, where they had visited Vesuvius and observed its productions, passed through Montelimart, a small town on the left bank of the Rhône, and, after dining with a party of *savans* resident there, amongst whom was M. Faujas de St. Fond, walked out to explore the neighbourhood. The pavement of the streets immediately attracted their attention. It is formed of short articulations of basaltic columns planted perpendicularly in the ground, and resembles in consequence those ancient roads in the vicinity of Rome, which are paved with polygonal slabs of lava. Upon inquiry they learnt that these stones were brought from the rock upon which the Castle of Rochemaure is built, on the opposite side of the Rhône; and were informed, moreover, that the mountains of the Vivarais abounded with similar rocks. This account determined the Academicians to visit that province, and step by step they reached the capital of Auvergne, discovering every day fresh reason to believe in the volcanic nature of the mountains they traversed. Here all doubts on the subject ceased. The currents of lava in the vicinity of Clermont, black and rugged as those of Vesuvius, descending uninterruptedly from some conical hills of scoriæ, most of which present a regular crater, convinced them of the truth of their conjectures, and they loudly proclaimed the interesting discovery.

On their return to Paris M. Guettard published a Memoir announcing the existence of volcanic remains in Auvergne,* but obtained very little credit. The idea appeared to most persons an extravagance; and even at Clermont a sagacious professor, who ascribed the volcanic scoriæ to the remains of iron-furnaces established in the neighbouring mountains by those authors of everything marvellous, the Romans, gained far more partisans than the naturalist. By degrees, however, the obstinacy of ignorance was forced to yield to conviction, and M. Desmarest some years afterwards, having published his ' Memoirs on the

---

* Mémoire sur quelques Montagnes de la France qui ont été des Volcans.—*Mém. de l'Acad. des Sciences*, 1752.

Origin of Basalt,'* accompanied with maps of many of the volcanic currents of Auvergne, put an end to all doubt upon the question.

Faujas de St. Fond, whose attention had by the circumstance above mentioned been directed towards the volcanic remains in his neighbourhood, published nearly at the same time his account 'Des Volcans éteints du Vivarais et Velay;'† but unfortunately, never having examined the phænomena of volcanos in activity, or learnt to distinguish with accuracy the substances which have been produced or altered by this class of natural agents, fell into numerous errors; mistook every chasm for a crater, every mass of basalt for a volcano, and saw nothing but decomposed lavas in beds of marl and sandstone. M. Desmarest's maps were astonishingly accurate, and evinced a very close and conscientious study of the localities. But his descriptive remarks contain many errors. He thought he saw, in every isolated fragment of an ancient basaltic current, what he called a *culot*, or residuum of lava stopping up the mouth of a crater; and he entirely neglected all observations on the mineralogical characters and distinctions of the various lavas he discovered. The labours, however, of both these naturalists were of the utmost service towards the establishment of the fact—that *numerous volcanos had broken forth in the interior of France at different and very remote periods*, and had covered most parts of the provinces of the Auvergne, Velay, and Vivarais, with the products of their eruptions.

Dolomieu likewise in the year 1797 passed hastily through Auvergne on his way to Switzerland; but in the report which upon his return he presented to the Institute he makes but very brief mention of its rocks, and appears only to have sought in

---

* Mémoires de l'Acad., 1771.   † Fol. 1778.

them a confirmation of his favourite doctrine, the igneous fluidity of the central globe.

Le Grand d'Aussy, in his 'Voyage en Auvergne,' published in 1794, again called the attention of the public to the natural phenomena of this country, which were sinking into neglect; but being more qualified for romantic descriptions than scientific observation, little real knowledge on the subject is to be gained from his work; and it was only in 1802 that M. de Montlosier, after a profound and attentive study of the Monts Dore and Dôme, published his Essay 'sur la Théorie des Volcans d'Auvergne,' and exhibited in their true light the various relations and distinct characters of these interesting volcanic remains. Desmarest, in his Memoir for 1771, had remarked that the perfect correspondence of repeated beds of basalt on each side of certain valleys demonstrates their having formed parts of the same currents, the continuity of which had only been interrupted by the excavation of the valleys subsequently to the flowing of the lava. Still he was far from applying this principle to its whole extent, and continued to look upon all the isolated hills capped by basalt, which are so numerous in the Auvergne, as so many volcanos denuded of their scoriæ by currents of the sea, under which he imagined them to have burst forth; and enveloped by those horizontal beds of limestone upon which in reality the volcanic products repose. M. de Montlosier then was the first to establish the true nature of these basaltic peaks and plateaux, and thus gave a key to the study of the country, without which its phenomena would present a suite of interminable difficulty. Indeed, the greater number of his remarks have been received as established facts by all later observers, whose subsequent researches have but confirmed their solidity.

M. de Buch, having paid a short visit to Clermont in 1802,

wrote in a letter to M. Pictet an account of some of the most striking volcanic remains in its vicinity, which was published immediately in the 'Bibliothèque Britannique,' together with a copy of a portion of Desmarest's map; and in 1809 he printed in the second volume of his 'Geognostichen Beobachtungen' some letters on Auvergne, accompanied by engravings of the Mont Dore, as an appendix to his 'Voyage en Italie,' &c.

In 1803 M. Lacoste, professor of Natural History at Clermont, published some observations on the Volcanos of Auvergne, and in 1805 some letters on the same subject. They contain, however, little information of value. In 1808 M. Ramond, then præfect of the department of the Puy de Dôme, read to the Institute his 'Mémoires sur le Nivellement des Plaines,' to which was subjoined a list of heights measured barometrically in the neighbourhood of Clermont, and a compendious account of their geological characters. A short time afterwards, M. d'Aubuisson, who subsequently published a general treatise on Geognosy, laid before the Institute his observations made during a tour through the Auvergne, Velay, and Vivarais; a tour, the result of which is well known to have been his speedy conversion from a thorough belief in the Neptunian origin of basalt and the other members of the flœtz trap formation, to an entire conviction of their having flowed in a state of igneous liquefaction, and their complete identity with volcanic lavas—a conclusion to which indeed no one could fail to arrive who would pursue the same straight road to the truth as M. d'Aubuisson.*

In 1815 M. le Baron Ramond presented to the Institute a second Mémoire on the subject, entitled 'Nivellement Barométrique des Monts Dors et des Monts Dôme,' which, besides the absolute heights of more than two hundred and fifty remark-

---

* Journal de Physique, tomes 58, 59.

able points upon these two groups of mountains, contains a detailed and methodical relation of their geological structure, drawn up with that scientific discernment and impartiality which have placed M. Ramond among the most distinguished observers of his age. To this work I have been much indebted, as well for a great body of local information as for nearly all the measurements of height I have made use of in the following pages.

These, with the exception of a paper of Dr. Daubeny's in the 'Edinburgh Philosophical Journal for 1820-21,' and a few scattered and partial notices inserted in the 'Journal des Mines,' by MM. le Coq, de Laizer, and Cordier, are the only writings I had met with on any of the volcanic remains of the interior of France at the time the first edition of this Memoir was published. Since that date the country has been again visited by Dr. Daubeny in 1830, and likewise by Messrs. Lyell and Murchison in 1829. The observations of the former are consigned to his volume on Volcanos (2nd ed., 1848). Those of the latter geologists were in part communicated to the public in the 'Journal of the Geological Society' (vol. ii. p. 75), together with a paper in the 'Edinburgh Phil. Journ.,' April, 1829. Sir C. Lyell has since briefly and most ably sketched the geology of Auvergne, &c., in his 'Manual.' References to the principal publications that have appeared on this subject from the French geologists are given below.* None, however, are of so complete

---

* M. Rozet, Mémoire sur les Volcans de l'Auvergne: Bulletin XIII., p. 221. MM. Le Coq et Bouillet, Vues, &c., du Département du Puy de Dôme, 1830. M. Fournet, Ann. des Soc. Géol., i. p. 225, 1842. M. Ruelle, Bull. XIV., p. 106, 1842. M. Raulin, Bull. XIV., p. 547. M. Burat, Description des Terrains Volcaniques de la France Centrale, Paris, 1843. M. Passis, Bull. XIV., p. 240, 1843. M. Pomel, Bull. XIV., p. 206, 1843. Id., 2de Sér., i. p. 579, 1844. M. Prevost, Bull. XIV., p. 125. M. Aymard, Bull. 2de Sér., vols. ii., iii., iv. M. Bertrand Roux, Description du Puy en Velay, 1823—an excellent work so far as it goes. See also the Appendix to Daubeny's Volcanos, 1848.

a character as to serve the purpose of a general description of this very remarkable country, or as a guide to geologists who may wish to examine for themselves in detail its singular features, more especially that series of volcanic products which show themselves there in a variety of combinations and positions of peculiar and perhaps elsewhere unexampled interest. It is hoped consequently that the present publication may be of some service in that capacity.

CHAPTER IV.

GENERAL ACCOUNT OF THE VOLCANIC FORMATIONS OCCURRING UPON THE ELEVATED GRANITIC PLATFORM OF CENTRAL FRANCE.

§ 1. It has been already mentioned that the volcanic formations of Central France attain an elevation much superior to that of the highest parts of the granitic platform. By many of their earliest observers, especially by M. de Montlosier, and subsequently by Dr. Daubeny, they were described as of two classes, ancient and modern, according as they appear to have been produced before or after some supposed epoch of a diluvial character, to which the excavation of the existing valleys of the district was attributed. My observations of the general features of the country in 1821 led me not merely to doubt, but to deny altogether, that there is any reason for referring the denudatory action to which its valleys are due to any single cataclysm or diluvial phenomenon. It appeared to me clear that this process has been going on from the first appearance of the land above the sea—that it is still in action, being chiefly occasioned by the decomposition and erosion of rocks by rain, frost, and other meteoric agents, but especially by the direct fall of rain from the sky, and the wash of the superficial waters, which everywhere and on every scale of force, as rills, streams, and rivers, are ceaselessly engaged in sapping and mining their banks, and carrying off the detritus to the plains, which they cover with alluvium. This every flood drives further, and grinds still finer, until it is ultimately carried out in the shape of sand or mud into the sea, where it settles as a sedimentary deposit. I thought I saw ample proofs in the relative position of the plateaux of basalt and trachyte which are seen capping so many of the hills in Auvergne at various elevations, some more than a thousand feet above the level of the plain of the Limagne, others but slightly raised above its surface, or the alluvial bottoms of its tributary valleys, that the excavation of these valleys, as well as of the plain into which they descend, has been gradual from the earliest to the latest times, and accompanied throughout by occasional volcanic eruptions, chiefly from the neighbouring granitic heights, but sometimes from within the area of the lake-basin, that is, of the existing plain of the Limagne; and consequently that no clear chronological line of separation can be drawn between the ancient and modern volcanic products; although, no doubt, some are of a very remote antiquity as compared with others.

It seemed to me, viewing as a whole the entire district of Auvergne, the Velay, and the Vivarais, that its volcanic rocks divided themselves geographically into six distinct groups, viz.— first, the three mountain masses of the Mont Dore, the Cantal, and the Mezen, each of which raises its colossal bulk from the granitic platform to a height of about six thousand feet above the sea, and appears to have been a centre of repeated eruptions on the largest scale, giving birth to a volcanic mountain like Etna, Teneriffe, and other sites of habitually recurring eruptions; 2ndly, the products of more isolated vents of eruption which broke out at various periods, but by far the greater number since the quiescence of the habitual volcanos above mentioned, upon a zone running nearly north-west and south-east from the north-west of Riom to the neighbourhood of Aubenas on the Ardèche. Some rather wider breaks than usual between the points of eruption on this line induced me to divide this group also into three sections: 1, the chain of puys of the Monts Dôme; 2, that of the Haute Loire; and 3, the cluster of volcanic vents of the Vivarais which have broken out in some tributary gorges of the Ardèche. To these indeed should be added a fourth and independent group, which I have not myself examined, and of which I have not succeeded in finding any detailed description. It occurs south of the Cantal, near La Guiole, and on a line parallel to the eastern zone already mentioned.

The geographical convenience of this division seems to have led to its adoption by later writers, both French and English: I shall therefore continue to employ it in the following pages. But for many reasons—especially that it forms the first in order of approach to visitors from Paris and the north, and the best introduction moreover to the other volcanic phenomena of the district—I shall begin with the description of the chain of puys (as they are called) of the Limagne and the Monts Dôme, which rise to the west of Clermont-Ferrand, the chief town of the department of the Puy de Dôme, and an excellent centre from which to explore the remarkable country around.

CHAPTER V.

FIRST VOLCANIC REGION.—MONTS DOME AND THE LIMAGNE D'AUVERGNE.

The Limagne d'Auvergne, as has been already noticed on the subject of its freshwater formation, is an extensive valley-plain, about twenty miles in breadth and forty in length; its soil, with the exception of the calcareous hills already described, being an alluvium consisting chiefly of boulders of granitic rocks, trachyte, and basalt, through which the Allier for the most part still wears its channel in a course from south to north. The inclination of the surface of the plain towards the river on either side, where not interrupted by hills, averages perhaps twenty feet in a mile. This at least is its slope from the base of the low hill on which Clermont stands, to the low-water mark at Pont du Château, the first point being 1204, the last 1027 English feet above the sea, and the distance about nine miles.

The western limit of the plain is formed by the abrupt escarpment of the granitic platform already described, which is fringed by some lower hills that branch off from it into the plain, and furrowed by steep and short ravines. These on being explored are found to penetrate to no great distance, terminating at the base of the range of volcanic hills, or "puys" as they are locally called, which rise from the otherwise nearly level plateau in a line nearly due north and south. On the western side of this chain of puys the platform slopes more gradually towards the valley of the Sioule, which runs nearly in a parallel direction. The width of this granitic table-land is about twelve miles; its average elevation 2800 feet, being about 1600 above the plain of the Limagne; but on some points where it has been preserved from denudation by a capping of basalt it attains 3300 feet. On its western side it is composed chiefly of gneiss, but on the east of veined granite, in which transitions from a coarse to an extremely fine grain are very frequent. Much of this rock readily decomposes, and every storm washes away heaps of crystalline sand from its exposed surfaces. Upon this platform rises the "chain of puys," comprehending about seventy volcanic hills of various sizes, several of them being grouped together in immediate contact, in other cases a considerable interval intervening; the whole, together with the scoriæ and volcanic ashes which cover the plain around and between them, forming a notched and irregular ridge directed north and south, about twenty miles in length by two in breadth.*

With the exception of five (among which is the Puy de Dôme itself, the loftiest and most prominent of these hills), all of them are volcanic cones of eruption,† apparently of very recent production. Their height is from 500 to 1000 feet above their base. They are generally clothed with a coarse herbage or heather; some few with thick forests of beech, once much more abundant. This covering, however, does not hinder an examina-

---

* See Map of the Monts Dôme, and the profile sketches, in Plates I. and II.

† A volcanic "cone of eruption" in its normal form, with a "crater" or cup-shaped hollow at its summit, is the result of the accumulation round the volcanic orifice or vent of the scoriæ and other fragmentary matters projected into the air by the series of explosive discharges of elastic vapour and gases which usually characterises an eruption. The fragments which fall back into the vent are, of course, thrown up again and again, and triturated into gravelly sand or fine ashes by the friction attendant on this violent process. Those which fall on the outside of the vent are heaped up there in a circular bank, the sides of which, both within and without, slope at an angle rarely exceeding 33°. And this bank, viewed externally, has of course the shape of a truncated cone, the crater being a hollow inverted cone contained within it.

tion of their composition, which is betrayed by frequent rents and scars in the turf. Many considerable portions seem to have been always bare of vegetation.

They appear entirely and uniformly composed of loose scoriæ, blocks of lava, lapillo, and puzzolana, with occasional fragments of domite and granite. Their form is more or less that of a truncated cone; the sides rising at an angle which oscillates about 30°. The crater is often perfect, and the hill must then be mounted to observe it. Frequently, however, it is broken down on the side whence the lava issued, evidently by the weight and impetus of this substance, when propelled from the volcanic orifice in a fluid state after the formation of the cone.

In some instances eruptions have taken place on different points so near each other that their ejected matters have mingled; and in place of a single cone, an irregular hill with two or three distinguishable craters, or a long-backed uneven ridge, has been the result. One, the Puy de Montchié, has four distinct craters, which, however, are not visible till the hill is mounted.

It is probable that almost every one of these numerous vents has furnished its stream of lava; but the bases of the cones being sometimes partially cultivated, or covered with forests, it is impossible in every instance to discover this; nor, for the same reasons, is it easy to trace to its source every visible current, and ascertain the precise crater which produced it.

The volcano evidently sometimes continued to eject scoriæ and ashes after the lava had ceased to flow—a circumstance often remarked in the eruptions of Ætna: in this case the immediate source of the stream of lava and its connection with the crater will have been concealed by these loose materials. Sometimes, it would seem (and this also is common to the eruptions of all the recent volcanos I have had the opportunity of witnessing), the lava has been produced by one orifice, while the aëriform jets issued from another, the latter presenting an intact and complete cone of scoriæ and fragments, the former a breached and imperfect one.

In general, however, the currents of lava are observed to issue directly either from the crater or foot of the cone, and thence to spread over a wide expanse of the neighbouring plateau, or fill the bottom of a valley to some distance. Their surface presents a succession of shapeless and bristling masses of scoriform rock, and offers to the imagination the idea of a black and stormy sea of viscid matter suddenly congealed at the moment of its wildest agitation.

These fields of lava, which are either wholly bare or but partially clothed with stunted brush-wood, are called "Cheires" in the patois of Auvergne.*

The lava has flowed either to the east or west according to the natural levels of the ground and the situation of the vent; and though perhaps a larger number of currents have directed themselves towards the Limagne, entering some of the steep valleys which communicate with it, the gentle slope by which the plateau descends on the west to the Sioule has caused those which were poured out on that side to cover a wider space, and exhibit themselves in consequence more conspicuously.†

A few eruptions have taken place, on either side, out of the principal line of puys. The most distant of these may be

---

* It is singular enough that the inhabitants of the fertile regions at the base of Ætna should call their lava rocks by a similar name, "Sciara." Borelli latinizes this, "Glarea;" but the term "Serra" (saw), applied to the jagged outline of a mountain range, seems to be the common stock whence both words are derived. The serra or sierra of the Spaniards has a corresponding signification; one of the basaltic platforms of Auvergne is still called and spelt La Serre.

† See Plate II., and the Map of the Monts Dôme.

observed near the village of Chalucet on the further bank of the Sioule. The Puys Graveneire, Channat, and La Bannière, rise upon a second line parallel to the first and most considerable one, and immediately on the margin of the high granite plateau, where it overhangs the plain of the Limagne.

The hills of the principal chain are on many points arranged in particular groups or systems, in each of which the different cones were apparently thrown up by eruptions either contemporary or immediately succeeding one another, from ramifications of the same inferior vent. Their lavas sometimes resemble each other very closely; and seem occasionally to have been poured forth from neighbouring orifices into a single bath or sea, in which they are completely mingled, as if once in a state of fluidity at the same moment. From hence, as from a common reservoir, a solitary current generally takes its rise, advancing to the still lower levels. It is known that many of the eruptions of modern volcanos, such as Etna or Vesuvius, produce several minor or "parasitic" cones of this kind, and several more or less distinct streams of lava from orifices consecutively opened as those first formed get "plugged up," or as the lava forces its way through new points of the eruptive fissure.

The hills composing these groups adhere to no general arrangement. In some instances they form a straight line touching each other by their bases, or uniting even more closely; in others, an irregular group.

Although all the cones of the chain of puys may be considered of recent formation, they are far from belonging to a single epoch. Everything attests their having been thrown up by eruptions succeeding one another often at distant periods. The different aspects of their lava-currents, some of which have yielded considerably to external decomposition, while the surface of others is still bare, harsh, and uninjured, would not perhaps *alone* prove a sufficient criterion of age, since the power of time in producing these effects on lavas varies with their varieties of mineral constitution, the more or less of iron, felspar, &c., which they contain. The considerable dilapidation of some cones, and the elevated position of their currents relatively to the surrounding soil, are less fallible signs of a superior antiquity, more particularly where they coincide, as they invariably do in this district, with the former class of marks.

Nothing, however, like an approximation to a knowledge of their positive ages can be expected, however interesting the elucidation of this question would be. All we know is, that, in spite of the very fresh aspect of many of them, their production must have been anterior to the earliest historical records of this locality, in which no mention is found of any volcanic eruptions.

In the middle of the line just described rises the celebrated Puy de Dôme, the Giant of the Chain, as it is called by Ramond; far superior in bulk and elevation to the numerous hills which stretch from its base towards the north and south. Its height above the sea is 4842 feet, and about 1600 feet above its base, the sides sloping at an angle of from 30° to 60°. It consists entirely of that variety of trachyte which, from a supposed peculiarity of mineral character, has been named *Domite*.

This mountain, with four or five other neighbouring hills of much less size, composed of the same rock, are so closely connected in situation with some of the volcanic cones of the chain of puys, in the centre of which they are found, as, notwithstanding their totally distinct structure and composition, to leave no doubt, in my mind at least, of their being connected likewise in origin with these cones, of their having been produced, in short, at the same time, and by a modification of the same volcanic agency.

They are scattered irregularly amongst the other puys near

the middle of the series, and are distinguishable from them at a distance by their whitish tint where the rock is exposed, and by their rounder contours.

With the exception of the Puy Chopine they consist entirely of that rock to which the most considerable of them has given the name of domite; each appearing as one enormous mass of this substance, in which it is not easy to discover traces of any definite structure; in some parts nearly compact, solid, and moderately hard; light, earthy, friable, or completely pulverulent, in others. The substance of one mountain differs but in accidental and insignificant characters from that of another. The colour of the rock is generally a greyish or brownish white; but on many points it has acquired various tints of red and yellow from the action of acid vapours. It absorbs moisture with the greatest avidity, and this action is accompanied by a hissing noise and a considerable disengagement of air-bubbles. Its texture is rudely granular; and on examination with a lens it appears to be an aggregation of imperfect microscopic crystals of glassy felspar, sprinkled with still smaller grains of augite, and occasional pallets of mica. These elements are partly separated from each other by minute pores, which render the substance rough to the touch, light, spongy, and bibulous. The larger imbedded minerals are, the glassy variety of felspar, generally cracked and mealy; mica in hexagonal or rhomboidal pallets, either bronzed or black; hornblende in acicular hexagonal foliated crystals, generally of a deep black, and highly lustrous along the planes of cleavage; and sphene; as well as specular and titaniferous iron in dispersed grains, blade-shaped laminæ, or regular octohedrons.

That these different substances crystallized nearly at the same period, would seem from the intimate manner in which their crystals are interwoven. Yet in general it appears that those of mica and hornblende were entirely formed before the complete crystallization of the felspar; for perfect crystals of both these minerals may be frequently seen enveloped or as if suspended in the centre of a large crystal of glassy felspar, and the cracks produced in the mica or hornblende are invariably penetrated by felspar. Very commonly the component crystals appear to have been broken, bent, or split, probably by the movement which the enveloping substance suffered when propelled upon the surface of the earth from the volcanic focus.

This rock is extremely liable to decomposition, which affects it often to the depth of some feet. Its parts are then disaggregated, it assumes an earthy aspect, and crumbles between the fingers; the crystals of felspar become carious, lose their lustre; and finally the whole mass is resolved into a meagre and ashy powder, in which the crystals of hornblende, mica, and octohedral iron are found uninjured.

The volcanic nature of domite has never been contested; and indeed it is sufficiently evidenced by the pumice-stones which accompany and are enclosed by it; the vitreous nature of its crystals of felspar; by its being porous, impregnated with muriatic acid, coated with sulphur, with sublimations of iron, &c., not to mention its similarity to the trachyte of the Mont Dore, which has still stronger proofs of a volcanic origin.

Surrounded and partly embraced by cones of scoriæ and lapilli, and remote from the rocks of the Mont Dore, which alone they resemble in substance, these hills have offered a perplexing problem to all the cursory visitors of Auvergne, and at the same time an ample field for conjecture to theorists. Hence the contradictory opinions that have been started on their origin. Desmarest considered the rock which composes them as a granite calcined *in situ* by a volcanic conflagration environing it;—Saussure, a petrosilex which had undergone the same extraordinary operation;—Dolomieu, Mossier, Montlosier, and De Buch, a granite triturated or liquefied by volcanic agency below the earth, and then by a sudden expansion of gas propelled through different apertures, over which it was consolidated in the form of huge bubbles.

Ramond objected to these hypotheses in their full extent, and argued that to imagine a mountain of the magnitude of the Puy de Dôme baked throughout upon the spot, is a stretch of imagination which can only be equalled by the supposition of its having swelled up like a bubble without bursting.

He denies, moreover, that each of the domitic hills is independent of the rest, the insulated production of a single and local operation, and asserts that this rock shows itself partially on many other points in the vicinity—a fact tending to show that these separate masses of the same substance were once, if they are not still, united, and constituted together a large bed, covering an extensive surface of the plateau, which succeeding volcanic eruptions and other mechanical injuries have in great part destroyed, concealed, and reduced to the seemingly insulated remnants which now alone show themselves. M. Ramond concludes that this bed is a ramification from the trachytic currents of the Mont Dore. M. d'Aubuisson also has professed the same opinion; so that the authorities on either side seem nearly balanced.

It is, I believe, now generally recognised that domite is but a variety of trachyte; the same rock, in all essentials, which constitutes the greater part of the Mont Dore and Cantal, the Euganean Hills, the Monti Cimini, and the Isles of Lipari and Ponza.* It is therefore wholly unnecessary, nor should we be

---

* *Vide* Brocchi, Cat. Rag. *passim.*—Brieslak, Institutions Géologiques, vol. iii., &c. &c.

warranted in the attempt, to account for its production by any other mode of formation than that which appears common to similar rocks in other places; unless such an explanation should be in this instance opposed by any manifest improbability.

There is every reason to conclude the trachytes of the Mont Dore and Cantal, as well as the clinkstone of the Mezen, to have been propelled from a volcanic orifice in a state of incomplete liquefaction—in short, as lavas—and to have followed the inclination of the ground they occupied, flowing in a manner differing only from that of basaltic lavas in proportion to their different consistence and very inferior fluidity, or the accidental circumstances which may have concurred to modify their disposition.

It is evident that, under similar circumstances of the surrounding levels and of propulsive force, the tendency of a mass of lava to quit the neighbourhood of the orifice from which it is emitted will be in exact proportion to its fluidity; and when the fluidity is at its *minimum*, it will accumulate immediately around the orifice; one layer of the half-congealed and inert substance spreading over that which preceded it, till the whole assumes the form of a dome or bell-shaped hillock perforated in the centre by the chimney or vent, through which fresh matter may continue to be expelled, but which will at the end remain closed by that last sent up. Now the variety of trachyte which composes the Puy de Dôme and the neighbouring domitic puys, consisting almost wholly of felspar, and therefore possessing the lowest possible specific gravity, and at the same time a very rude and coarse grain and highly porous structure, is precisely that species of lava which we should expect à *priori* to have possessed the *minimum* of fluidity when protruded into the air; *

---

* See Considerations on Volcanos, p. 92-96.

and we therefore can understand perfectly why, instead of flowing in thin and continuous sheets or streams to a distance from its vent, like the basaltic lavas produced about the same time and from the same fissure, it has accumulated in dome and bell-shaped hillocks on the point where it was emitted. That this *was* the mode of production of these masses of trachyte, certainly that they were thrown up on the spots they now occupy, seems to me proved by their rising in every instance either from the middle or the side of a regular crater and cone of scoriæ.*

If it could be imagined possible for the volcano of the Mont Dore to have sent forth a vast current of trachyte in this direction, of which these hills have been supposed the remaining segments, in spite of the fact that the great elevation of the granite ridge upon which they rest above the surrounding country renders it the last of all directions which such a current could have taken, and in spite of the improbability that a rocky bed of which the Puy de Dôme, a mass rising 1600 feet above its base, is *merely a detached remnant*, should have left no traces of its existence in the interval between that mountain and the Mont Dore, a distance of 7 or 8 miles; yet a still stronger objection to this hypothesis remains behind, viz. the improbability that the position of *each* of these fragments should severally coincide exactly with that of the vent of a separate recent eruption; that the *only* points on which any considerable remnants of this supposed bed are to be found should be *precisely*

---

* See Plate III., and the map of the Monts Dôme. In Iceland M. Robert describes ('Voyage en Islande,' Paris, 1840) the Mont Baula as a pyramidal dome of yellowish-white very porous trachyte, partly columnar, at the foot of which is seen a contiguous crater of eruption. The resemblance of this to the Puy de Dôme, rising from the crater of the Petit Puy, is complete. In Hungary also, and elsewhere, trachytic domes are described as rising in the middle of crateriform hollows.

those on which, from the disturbance occasioned by the volcanic explosions, there is good reason to suppose it would have been destroyed and carried off.

The theory of Von Buch evidently approaches more nearly to this explanation than that of d'Aubuisson and Ramond. Von Buch supposes the domite to have been "granite liquefied by volcanic heat:" I also conceive it to have been, like all lavas, a mass of granite, or some congenerous crystalline matter, which, while confined beneath the superficial rocks at an intense temperature, was suddenly allowed to expand, by the giving way of the overlying rocks, and was consequently liquefied, or so far softened by the immediate generation of highly elastic fluids, both gases and aqueous vapour, through every part of its texture, as to be protruded, by the tumefaction incident to this process, through the clefts of the crust above.

The part of Von Buch's theory with which I agree the least is his supposing these hills to be hollow, and blown up like a bladder. I imagine, on the contrary, the aëriform and highly elastic fluids, the expansion of which elevated the lava, to have remained chiefly where they were generated, viz. in a state of uniform and intimate dissemination throughout the texture and between the crystalline particles of the porous and elastic mass; and not by any means to have united into one great bubble or dome beneath an overlying crust of the lava, as is implied in Von Buch's theory—a theory which Humboldt has adopted and applied, with some rashness, to all trachytic formations. I need not dwell any longer on this topic, as the laws by which I conceive trachytic, as well as other lavas, to have been determined in their rise and appearance on the surface of the earth, have been developed in detail in another work ('Considerations on Volcanos,' pp. 85-129); and that the peculiarities which distinguish these particular trachytic formations will be best noticed when they are individually described among the chain of puys of which they form links.

To this I proceed immediately, commencing their description with the Puy de Dôme, which rises in the middle of the chain, and taking the remaining hills in the order of their succession from this towards the north and south, pointing out at the same time the currents of basalt, or the domes of trachyte, which appear to have been produced from the several vents marked by the cones of ejected matters. The accompanying map* is referred to in the numerals prefixed to the names of these hills. It was drawn up entirely from my own observations, upon the basis of the old Government survey of Cassini, and I believe its accuracy may be relied on.†

Description of the Chain of Puys.

(62.) *Puy de Dôme.* (Absolute height 4844 feet.)—This mountain, which occupies the centre of the chain of puys, and in all its proportions leaves the others far behind, rises to the height of 1700 feet from the average level of the plateau around.

Its figure is that of an irregular cone very obtusely truncated, the upper surface sloping considerably to the west. A rounded eminence on its eastern summit gives it, when viewed from that side, somewhat of the shape of a cupola. Its appellation, however (in Latin Dumum or Podium Dumense), has no reference to this resemblance, but is derived from the woods with which it was formerly clothed, and the remains of which still cover the foot of its eastern slope. In consequence of the very different

---

* See map of the Monts Dôme.

† When I drew this map I was not acquainted with that of Desmarest, which might have assisted me; but on comparing the two I have found the latter generally to confirm my own observations, with the exception already noted, that Desmarest insisted on seeing a "culot" or dyke of eruption in every detached fragment of a basalt current.

consistence of its substance, which resists decomposition and atmospheric erosion much more in some parts than in others, two sides are, as well as the summit, entirely covered with thick grass, and present a smooth and gently curved outline, while that of the others is broken by rocky projections, giving them an aspect of a ruder character. The fissures of the harder portions of rock are lined both with laminar and octohedral crystals of specular iron; those of the more friable and earthy with similar sublimations, which in these points assume the brightest tints of blue and green, and occasionally with a coating of sulphur. The substance of the rock itself is in many places frequently stained throughout with varying shades of an ochry yellow, citron, rose, scarlet, and greyish-blue colour. Fragments of domite thus tinged, when rubbed together, give out a strong smell of muriatic acid, and M. Vauquelin has detected in them the presence of this acid by analysis. The exhalations to which these effects must be attributed were probably evolved through crevices from the interior of the mass of trachyte during the process of cooling, and were the result of the partial decomposition of its inner substance by intense heat, as the elastic vapour it contained escaped through the pores of its outward surface, and thus diminished the pressure which had before opposed the decomposing influence of heat.*

A chapel was anciently erected on the summit of the Puy de Dôme, of materials brought from a distance (the additional labour and expense incurring of course a redundancy of gratitude from Our Lady, to whom it was dedicated); and the ruins of this building have probably strewed the sides of the mountain with the blocks of basalt which are occasionally found there. The scoriæ occurring in the same situation have been evidently

---

* See Considerations on Volcanos, p. 125-6.

projected there by the explosions of the neighbouring cones. Exceedingly light and porous fragments of domite, resembling pumice in these characters and in their rudeness to the touch, but without its very vitreous filaments, also form part of the loose débris of this mountain, and many of them, appearing to have been rounded by attrition, may be observed imbedded in the more solid rock.

It is difficult to discover anything in the *structure* of the Puy de Dôme to indicate the particular mode of its production. I have given above the general conclusions to which I have arrived with reference to this as well as the other trachytic hills of the chain. On the south side the base of this mountain has been perforated by volcanic explosions, which in all probability immediately succeeded the protrusion of the trachytic lava, and which have left a large crater, half-encircled by abrupt cliffs of trachyte, and containing within it a complete cone composed of fragmentary trachyte, scoriæ, and lapillo. This cone is called the Puy de Besace (39), and the ridge of the surrounding crater the Puy de Gromanaux (38). On the north side of the Puy de Dôme, and immediately contiguous to it, is another crater belonging to the Petit Puy de Dôme, to the description of which we proceed.

Chain of Puys north of the Puy de Dôme.

1.* *Petit Puy de Dôme.*—A volcanic cone, one of the largest though not the most perfect of the chain, leans against the northern flank of the Puy de Dôme. Viewed from a distance, it appears to be a dependence of this latter mountain, and hence the name it has received of Le Petit Puy de Dôme. Upon a closer inspection the different nature of the materials of the two

---

* The numbers refer to those on the map of the chain.

mountains is immediately apparent; the Petit Dôme being composed entirely of fragmentary matter, basaltic scoriæ, sand, and ashes. This volcanic cone reaches an elevation of 4186 feet, and is lower by 656 feet than its colossal neighbour.

It contains a very regular crater, shaped like the interior of a deep bowl, and called by the mountain herdsmen "Le Nid de la Poule," "The Hen's Nest." Its diameter and depth are nearly equal; the latter measures 300 feet from the highest point of the circumference, of which the southern portion is much the lowest. To the north and east an outer semicircular ridge, running parallel to the inner one, appears to be the rim of an older crater defaced when that now existing was formed.

From the western base of this hill a stream of basaltic lava called the *Cheire de l'Aumône* takes its rise, and extends to a considerable distance over the slope of the granitic platform, but is at length lost under a cultivated soil, or confounded with other and seemingly later currents, which have taken a similar direction. This lava was most probably emitted at the epoch of the eruption of the Petit Dôme.

2. *The Great Suchet*—a saddle-shaped eminence with some slight traces of a crater on its southern slope—is connected with that last described by means of the *Little Suchet* (63), one of the domitic hills, which stands at the point of a right angle formed by the ridges of the hills it unites. The fragmentary matters that compose the Great Suchet were apparently projected by the same eruption which produced that mass of trachytic lava; and this idea, suggested by their proximity in position, is confirmed by the numerous blocks of domite and pieces of pumice which enter into its composition.

3. *Puy de Côme.*—Rather to the west of the meridian on which stand most of the puys, rises that of Côme, remarkable for the regularity of its conical form; for its height, exceeding 900

feet from the plain around, whence it rises majestically at an angle of about 35°; and, above all, for the prodigious dimensions of the lava-current it has poured forth. This current is also one of the most interesting of the whole chain, from the obstacles it has met with in its course, and the changes it has effected in the surface of the ground it overflowed. The sides of this hill are covered with forest-wood; its summit presents two distinct and very regular craters; one of them, with a vertical depth of 250 feet, is of considerable diameter.

The cone was apparently thrown up after or during the eruption of its stream of lava; which, instead of issuing from either crater, takes its rise at the western base of the hill.

At no great distance from its source the lava encountered an angular protuberance of granite, which evidently caused the current to separate into two branches. That to the right, the most considerable of the two, spread itself over a vast surface, and, aided by the gradual inclination of the granite platform, pursued its course to the west till it found an obstacle in a long line of hill, consisting of alluvial tuff from the Mont Dore covered by an ancient plateau of basalt. Thus impeded in its progress, the lava followed the sweep of the hill in a north-east direction; and finding an issue at length between this and a granitic knoll which obstructed its progress towards the north, poured down on the present site of the castle and town of Pont Gibaud; immediately above which it seems to have met and flowed over a more ancient stream from the Puy de Louchadière (No. 21).

Thence both together poured in a broad sheet of lava down the steep side of a granitic hill which formed the eastern border of the valley of the Sioule, dashing forward against the rocks on the opposite side, and usurping the channel of the river, down which they pursued their course to the distance of more than a mile.

The Sioule, thus dispossessed of its bed, has been constrained to work out a fresh one between the lava and the granite of its western bank, which in consequence is extremely precipitous. But, before this was accomplished, there is every appearance of its waters having been so far obstructed as to create a lake, covering the flat and alluvial surface now forming the meadows of Pont Gibaud.

3.   Rock of Prismatic Lava on the Banks of the Sioule, near Pont Gibaud.

In one part of this new channel, where the valley has a slight bend, and the torrent of lava being opposed by a salient rock accumulated to a considerable thickness, the excavation effected by the river has disclosed its internal division into vertical jointed columns, the lower portions of which are straight and well formed, the upper twisted into various curves, and less regularly polygonal. The wall of lava is about fifty feet high, and the columnar division is prolonged incompletely to the extent of between 200 and 300 yards. (See the woodcut overleaf.)

The rest of the lava filling the valley where it is about the third of a mile in breadth, is split by fissures, mostly vertical, into amorphous masses which still at various points evince a tendency to the prismatic form. This is perhaps the most marked instance of all Auvergne in which the lava of one of the very recent volcanos assumes a decided columnar division. In the Vivarais, as will be seen in a future chapter, this circumstance is of frequent occurrence.

The sketch (Plate II.) was taken from a basaltic plateau on the summit of the granitic range of hills, west of the Sioule, from whence the greater number of puys may be observed; that of Côme being the most conspicuous as well as the nearest. The whole course of this branch of its lava, as well as that of the Puy de Louchadière, is seen from this point.

But the Sioule was not to suffer from this invasion alone. The other branch of the lava-current of Côme, called the Cheire de l'Aumône, which flowed on from the point of separation in a direction west-south-west, soon reached the bed of this river, about three miles above the spot of the other irruption, and pouring over its banks filled up the entire valley with an immense causeway more than 100 feet high. Exhausted by this effort, it proceeded but a short way down the bed of the stream towards the north, and stopped where the village of Mazayes now stands.

The baffled waters of the Sioule here, as at Pont Gibaud, obstructed by the rocky dyke thus suddenly thrown across their channel, must have given birth to a lake by their stagnation, and would probably have ended, as in the other instance, by wearing away a passage parallel to their former one, had not the hill forming their western bank, not in this instance composed of granite, but of a soft alluvial tuff, yielded, at some distance up the stream, to the excessive pressure of the dammed-up waters. An immense excavation, still subsisting, was broken across this hill, through which the lake emptied itself into the bed of the Monges at no great distance, and through which the Sioule still joins this latter stream about three miles above their former confluence.

A considerable body of back-water still remained behind in the part of the valley of the Sioule intercepted between the dyke of lava and the emissory thus forcibly created, which, from the inclination of the ground to the north, could not vent itself by this opening; and here a stagnant piece of water called the Etang de Fung, and used by the Seigneurs of Pont Gibaud as a stock-pond, existed till within a few years, when its drainage was artificially effected. The high banks which rise on each side of the long marshy meadow now occupying the site of the old pond still present the correspondence of angles so generally observable along channels of running water.

On the opposite side of the enormous causeway of lava to which this diversion is owing, another small piece of water, occupying the bed of a tributary rivulet choked up in the same manner, is dignified by the name of the Lake of Mazayes, and its insignificant drain now runs into the Sioule through the remainder of the wide and deep valley which this river itself once excavated and possessed.

The changes thus effected do not only present themselves to

the eye of a nice observer, but are exhibited in a manner not to be mistaken by the most casual; and in fact they result so simply and necessarily from the causes brought into action, that I should not have dwelt so long upon their details, but that they serve to exemplify the mode of formation of other lakes in the Auvergne and Velay, and the origin of other changes in the surface of a country or direction of its rivers, where every link in the chain of causes and effects is not quite so palpable.

The whole superficies of the plateau covered by the lava of Côme cannot be estimated under ten square miles. Its thickness is not to be ascertained, since no excavation has been made through it; where the current has met with any obstacle, it must of necessity be considerable, and thirty feet may be taken as the probable average.

It is one of the most rugged Cheires of the Monts Dôme, presenting a succession of continual asperities, following one another like the waves of an ocean, with similar depressions between. Upon walking over its surface,—no easy task,—it appears to consist of chaotic heaps of rocky and angular blocks of compact basalt, tossed together in every variety of disorder; yet, in the deep and narrow intervals between these heaps, occur little patches of fresh and flowery turf, and knots of underwood spring from their clefts, contrasting strangely with the horrid desolation which prevails over this extensive wilderness.

Near the limit of the northern current, at some distance from Pont Gibaud, is a natural grotto in the basalt; from its interior gushes a small spring, which is partly frozen during the greatest heats of summer, and is said to be warm in winter; probably, however, only seeming warm by contrast with the external temperature.

The water is apparently frozen by means of the powerful

evaporation produced by a current of very dry air issuing from some long fissures or arched galleries which communicate with the cave, and owing its dryness to the absorbent qualities of the lava through which it passes. This is a phenomenon common to caverns in other volcanic districts.

The basalt of the Cheires de Come and de l'Aumône is almost identical in its mineralogical characters; it is replete with irregular cellular cavities, is of a dark greyish-blue colour, and contains a notable quantity of felspar. It is sufficiently tough to be worked for building-stone, and cuts well under the chisel. Nodules of pure white quartz are not unfrequently found enveloped by it, appearing to have suffered from heat only on the surface, which is cracked and sometimes shows a commencement of fusion.

4, 5. *Puys de Barmet and Filhou.*—Two small cones, the last of which has a semi-crater well marked out; that of the former is nearly effaced. These two hillocks were probably thrown up from side apertures at the epoch of the eruption of the Puy de Côme.

6. *Puy de Pariou.*—From its aspect, and that of its lava, this puy may be supposed the product of one of the last eruptions which convulsed the country. It is also one of the most considerable and regular cones of the chain.

A segment of an older crater half encircles it on the north, and here the process by which this and similar hills have been produced is manifest. The aëriform fluids developed in the interior of an ebullient body of subterranean lava, having by their expansive force effected a fracture in the overlying rocks, struggle upwards in powerful bubbles through this aperture, carrying with them at every eructation showers of scoriæ and their comminuted fragments. These accumulate round the margin of the orifice into a hill, necessarily approaching more or less in figure to the solid which would be generated by the revolution of an obtuse scalene triangle round one of the angles of its base; and this hill is called a volcanic cone, which, it must not be forgotten, when most perfect, is but a truncation of the geometrical one.*

But when the intumescent lava itself has subsequently risen and poured itself forth from the same vent, it must by its weight and impetus generally break down and carry away part of the loosely built hill encircling the aperture, and a semi-cone only then remains, the crater appearing open on one side. If the lava escapes from some other orifice, or by some channel beneath the foot of the cone, and does not rise within the crater, a complete cone is the result, having a perfect crater at its summit.

After the lava has ceased to be emitted, or has established another channel for its efflux, the gaseous explosions still usually continue for a certain time, by which a second cone is formed. This in the absence of subsequent disturbing causes remains perfect, as has happened in the example of the Puy de Pariou now under consideration, where since the first outburst of its vast current of lava the volcanic mouth evidently threw up prodigious showers of scoriæ and puzzolana, creating the beautiful cone and crater of Pariou upon the brink of that which had been formed by the earlier explosions of the eruption.

This newest crater has the figure of an inverted cone. It is clothed to the bottom with grass, and it is a somewhat singular spectacle to see a herd of cattle quietly grazing above the orifice whence such furious explosions once broke forth. Their tracks round the shelving sides of the basin in steps rising one above the other, like the seats of an amphitheatre, make the

---

* See note to p. 41.

excessive regularity of its circular basin more remarkable to the eye.

Its depth is 300 feet, and circumference 3000. The inclination of the sides of the cone and of the crater are both about 35°. The acute ridge resulting from their junction is so little blunted by time, that in some parts it scarcely affords space to stand on. Its highest point is 738 feet above the southern base of the puy.

The lava of Pariou is as instructive with regard to the circumstances which accompany the movement of this substance in viscid torrents over a large tract of country, as its cone on those by which such mountainous excrescences are suddenly thrown up.

Its first direction is to the north-east, and the current appears to have set strongly against a long-backed granitic eminence opposing it on that side. Thence, led by a considerable slope towards the south-east, it coasted the base of this hill; and leaving to the right another protuberance of the primitive plateau on which now stand the church and hamlet of Orcines, advanced to a spot called La Baraque. Here it met with a small knoll of granite capped with scoriæ and volcanic bombs, marking the source of a much more ancient basaltic bed known by the name of Prudelles. Impeded in its progress, the lava accumulated on this point into a long and elevated ridge, which still bears the appearance of a huge wave about to break over the seemingly insignificant obstacle. But an easier issue offered itself in two lateral valleys having their origin in the part of the plateau occupied by the lava-current; which, separating consequently into two branches, rushed down the declivities presented on either side.

The right-hand branch first deluged and completely filled an area surrounded by granitic eminences, and probably the basin of a small lake; thence entered the valley of Villar, a steep and sinuous gorge, which it threaded exactly in the manner of a watery torrent, turning all the projecting rocks, dashing in cascades through the narrowest parts, and widening its current where the space permitted, till, on reaching the embouchure of the valley in the great plain of the Limagne, it stopped at a spot called Fontmore; where its termination constitutes a rock about 50 feet high, now quarried for building-stone. From the base of this rock gushes a plentiful spring, the waters of which still find their way from Villar beneath the lava which usurped their ancient channel.*

The branch which separated to the left plunged down a steep bank into the valley of Gresinier, replacing the rivulet that flowed there with a black and shagged current of lava; entered the limits of the Limagne at the village of Durtol; and continuing the course marked out by the streamlet, turned to the north, occupied the bottom of the valley lying between the calcareous mountain Les Côtes and the curtain of granitic rocks, and finally stopped on the site of the village of Nohanent. Here, as at Fontmore, an abundance of the purest water springs from below the extremity of the lava-current. The various rills which drain the valley of Durtol and its embranchments, have recovered their pristine channel, and, filtering through the scoriform masses which always form the lowest surface of a bed of lava, flow on unseen, till the rock above terminates, and they issue in a full and brilliant spring. Above this point, consequently, is seen the anomaly of a valley without any visible stream; and the inhabitants of Durtol are condemned in seasons of drought to the strange necessity of seeking at Nohanent, a distance of two miles, the water which flows there beneath their

---

* These and the other lava-currents of the chain of puys are all to be traced on the map of the Monts Dôme.

own houses. A similar phenomenon is very general throughout the Auvergne, wherever a current of recent lava has occupied the bed of a mountain rivulet, not sufficiently copious or violent to undermine the lava above, or open a new side channel through its former banks.

In its appearance the Cheire of Pariou is even more bristling and rugged than those already described. M. d'Aubuisson justly compares it to a river suddenly frozen over by the stoppage and union of immense fragments of drift ice.

In the work to which I have occasionally referred on the general laws and conduct of the volcanic forces,* I suggested that this asperity of surface in a lava-current is probably owing, and appears usually proportioned, to the high specific gravity of the lava, which determines the ascending force of the bubbles of vapour that, expanding in its interior, rise, and escape from the surface as it flows onward. The lava of Pariou, which is composed almost wholly of augite, and therefore of high specific gravity, confirms this law. The same circumstance accounts for the other characteristics of this rock, which is perfectly compact in the interior of the current, while the outer portions are cellular and cavernous. The cavities, often of large dimensions, are lined with a dark vitreous varnish, and from their sides project numerous stalactitic protuberances, and slender filaments, coated in the same manner.

The colour of this basalt is a deep shade of bluish-grey; it contains imbedded crystals of augite, and a few of glassy felspar, and bears a great resemblance on the whole to the current from the Monti Rossi, which in 1669 destroyed a part of Catania, and reached the sea.

* Considerations on Volcanos. London: Phillips, 1825, pp. 119-20.

It was in all probability the eruption of the Puy de Pariou which threw up some large accumulations of puzzolana that are observable on both sides of the granitic ridge of Ternant and Clersat.

7. *Puy de Fraisse.*—Between the Puys of Pariou and Côme lies that of Fraisse, a saddle-shaped hill, devoid of any peculiar interest, except that its basaltic lava, which is seen only in erratic blocks, differs from that of the Puys already described by containing crystals of olivine.

64. *Cliersou.*—This puy occupies the centre of an area formed by the four volcanic cones, Pariou, Le Grand Suchet, Côme, and La Fraisse. Its figure is most precisely that of a bell, as the engraving will show.* Turf and brushwood cover the swelling curve of its steep sides, and a ring of broken rocks forms the edge of the circular and flattish capping which crowns it. This capping appears to be remains of an outer envelope, the lower parts of which have been worn away. The whole substance of the hill is a light and porous trachyte, differing in nothing from some parts of the Puy de Dôme. The upper half of the hill is perforated in all directions by caves and galleries formerly pierced for the extraction of this stone, which is supposed to have been held in esteem by the Roman colonists as a material for sarcophagi, in consequence of its properties of absorption. These caves abound in fragments of pumice detached from their walls, in the substance of which much of this mineral is observed to be enveloped. Pumice we know stands in the same relation to trachytic rocks as scoriæ to basalt, and these pumiciform parts are created of course in the former rock by the same local and accidental development of gases which occasion the

---

* See Plate III.

extremely cellular and scoriform parts often found in the substance of the latter.

M. Ramond supposes Cliersou and Le Petit Suchet to have been united formerly; and indeed this follows of course from his theory as to the origin of all these domitic hills from the Mont Dore. I am clearly of opinion that it was protruded from a vent on the spot where it now stands, and that it owes its form, as explained above, to the very low fluidity of its substance at the time of its emission. If any of the surrounding cones of scoriæ was thrown up by the same eruption, it must most probably be the Puy de la Fraisse.

8, 65, 9. *Puys de Sarcouï* and *des Goules.*—Immediately to the north of Pariou rises a linear group consisting of three hills and having the same direction with the general chain, viz. north and south. The central one, Le Grand Sarcouï, consists entirely of domite; the two others, Le Puy des Goules and Le Petit Sarcouï, are ordinary volcanic cones.

The first of these latter hills has a shallow but wide crater at its summit. It is remarkable only for the fragments, of gneiss apparently much altered by volcanic heat, of a black trachyte passing into a resinous state and enveloping pieces of gneiss, and of a variety of basalt with variolitic spots, which occur mingled with scoriæ, puzzolana, and ashes, in its composition. This has been exposed by the cutting of the high road from Clermont to Limoges, which occupies a narrow pass between the Puy des Goules and that of Pariou at an absolute elevation of 3310 feet, and during the winter is subject to dangerous *tourmentes*, or snow-storms.

The Little Sarcouï has a semi-crater, exactly fronting and embracing part of the circumference of its larger namesake. It gives rise to a current of lava on its eastern side, which follows the sweep of a granitic ridge as far as the village of Egaules, where it is confounded with that of the Puy de Jume. In mineralogical characters it differs little from that of Pariou.

65. *Le Grand Sarcoui.*—Between the two last described hills, and in close contact with each at its base, rises the Puy de Sarcoüi, one of the most remarkable and perfect of the trachytic domes of this district. In figure it is completely a flattened and rather elongated hemisphere, and is aptly compared by the mountain shepherds to a kettle placed bottom upwards. Of this singular shape, which attracts the attention from a great distance, the engraving (Plate III.) will convey a very accurate idea. The trachyte which composes this hill is rather looser in texture and lighter in colour than that of the Puy de Dôme; it has also fewer imbedded crystals, but is essentially the same rock. Like the Puy de Dôme, Sarcoüi has parts of its substance impregnated with muriatic acid, and stained with mixed and brilliant hues of red and yellow. Like Cliersou it has been quarried for sarcophagi, some of which still remain in an unfinished state in its caves, and the name of the hill is supposed to be derived from the purpose to which its stone was applied. In these caves I thought I perceived indications of the structure of the hill in massive beds, which, as far as can be made out from the thick vegetation and quantity of débris which cover the surface, follow the curving slope of its sides, and envelop its nucleus concentrically, in the manner of the coats of an onion. It has been already remarked that the Puy de Cliersou appears to show traces of a similar concentric structure.

The summit is nearly circular and quite flattened, covered thickly with heather, and strewed with scoriæ launched probably from some of the neighbouring puys. Near the foot of the hill on the east a number of small rocky knolls of trachyte pierce through the cultivated soil, and appear to connect themselves with the base of this larger mass of the same substance.

The very intimate union of Sarcouï with the Puy des Goules, together with the fact that the latter cone does not appear connected with any current of basalt, and moreover consists in great part of fragmentary trachyte, leads to the opinion that these two hills are of contemporaneous production, the latter being formed of the fragmentary substances projected by aëriform explosions accompanying or rather preceding the emission of the trachytic lava of Sarcouï.

This group of volcanic hills deserves to be thoroughly studied by those who wish to obtain an insight into the mode of formation of the dome-shaped masses of trachyte to be met with in many other volcanic districts, and which sometimes attain to the magnitude of the colossal Chimborazo. A rock of trachyte is rarely to be observed on so small a scale, so completely insulated from every non-volcanic rock in position, so accessible to examination, or presenting so perfectly the figure which has been supposed characteristic of this formation. Those who will imagine the effect which would be produced by heating a very thick soufflet-pudding in a closely covered vessel which it completely fills, until its intumescence force it to exude through a crack or hole in the cover of the vessel, over which the matter, quickly congealing by exposure to the air, would cake into a bulky excrescence, will understand exactly the mode of formation which I attribute to the trachytic dome of Sarcouï, the Puy de Dôme, and the other similar hills. Their substance, indeed, in its spongy and porous texture, and lightness, is really not very dissimilar from the sort of cake which would be the result of the homely culinary process I have here imagined. I beg pardon for so mean an illustration, but it explains my theory of the formation of domite better than any elaborate description. I must add, that on revisiting these hills in the present year I found no reason to alter the opinion I had formed of their origin in 1821.*

10. *Le Creux Morel.*—To the north of the Puy de Fraisse, and west of the group last described, is a large extent of level plain, covered with turf; the soil is a medley of volcanic fragments ejected by the surrounding cones.

A singular depression is observed in it near the Puy des Goules, evidently the result of a volcanic explosion, but differing from other craters in having a conical mount of scoriæ raised on one side only, while the margin of its remaining circumference is even with the plain around. It is 115 feet deep, has no lava-current, and was probably but a subsidiary spiracle of one of the neighbouring volcanic vents. A violent wind, or a slanting

---

* M. le Coq, though believing that they were formed at the time of the eruption of the neighbouring cones of scoriæ and basaltic lavas, adheres still, I believe, to the notion that these domitic puys are all hollow bubbles, blown up by elastic vapours from a fused portion of a pre-existing bed of domite or trachytic conglomerate. Although I think it perhaps possible that bubbles of such a size may have been formed in some *subterranean* lava masses, it would seem that domite, from its light, granular, and porous character, is the very last kind of lava in which they may be expected to occur. It is in the fine-grained, compact, and almost vitreous lavas that the largest bubbles are formed. There are indeed no bubbles or vesicles at all to be seen in domite. It is open and loose in texture throughout, as contra-distinguished from the compact basaltic lavas, in the superficial and scoriform parts of which bubbles are frequent. Some of the Icelandic lava-currents contain very large vesicular caverns, but they occur in the basaltic lavas, not in the trachytic ones. In the Mont Dore, as will be seen hereafter, many of its trachytic lavas take the form of vast massive hummocks, their imperfect fluidity having prevented their flowing to any distance from the vent whence they were protruded. The domitic hills of the Monts Dôme are in my opinion only masses of the same kind, of which the substance was when erupted in a still less fluid condition, intumescent throughout, but having the elastic fluids which gave it that character disseminated throughout the mass: whence its porosity and lightness. If it were considered worth while to solve this question by positive examination, it would not be a difficult or very expensive operation to drive a gallery a few hundred feet into the side of Sarcouï, even to its very centre. Those who think a vast empty space would be found there would, I imagine, be then convinced of their error.

line of projection, may have caused the matters thrown up to fall upon the side rather than around the orifice.

Its scoriæ contain crystals of black and green augite, and of olivine; the erratic blocks of basalt are unusually dense and hard.

11. *Puy de Chaumont.*—This is a very regular and large cone, with a crater at the summit nearly effaced by the accumulation of débris. Its lava and scoriæ resemble those of the Creux Morel, with which it was probably contemporaneous. No perceivable current; but the basaltic rocks which here and there pierce through the soil of the plain on the east, are rather referable to this than any other puy.

12. *The Puy de Lantegy* is scarcely more than a hillock of scoriæ, covered by numerous fragments of domite, with a small crater broken down towards the west. Its lava, like that of the other puys in the vicinity, is soon lost under the thick bed of volcanic ashes sprinkled over the plain, and the occasional cultivation which this circumstance induces.

13. *Puy de la Goutte.*—This is a segment of an immense crater connected in situation, and doubtless in origin, with the extraordinary medley of rocks called Puy Chopine, which it half encircles.

Fresh-looking scoriæ, and blocks of a compact basalt containing olivine and augite, and fragments of domite, show themselves on its sides and interior, wherever the turf is rent away, and the elevation of the soil to the west seems to indicate a current to have taken that direction.

14. *Puy de Leironne.*—This cone rises on the north of Chopine, that is, the opposite side from La Goutte, and by the two it is almost completely enclosed. Leironne has a shallow crater on the side towards Chopine, and seems to be composed more of domitic fragments, often scorified and approaching to pumice,

than of augitic scoriæ. It was probably coeval in origin with Chopine and La Goutte.

66. *The Puy Chopine.*—This is a very remarkable volcanic production, and has proved, and still will prove, a most perplexing subject of study to every geologist who visits Auvergne. D'Aubuisson left it after three visits, and candidly avows that he had not acquired any light on its probable mode of formation, nor any positive idea of its structure and composition. It may therefore appear presumption even to attempt to describe it; but, though to obtain a clear knowledge of the composition of a hill which is evidently a confused medley of heterogeneous substances is at the outset impossible, yet its general aspect and principal features may be without difficulty remarked and sketched.

The figure of the puy is irregularly conical, and its sides on some points are so steep that vegetation is prevented from covering them by the fragments which are continually detaching themselves and rolling to the bottom; and hence two wide rents on the east and south lay bare its upper half; the lower is concealed by the immense and daily increasing accumulations of débris.

Its composition, as far as it can be observed by help of these barren faces, which from their excessive steepness and the crumbling nature of their materials are very difficult of access, is a mixed assemblage of various primary crystalline and volcanic rocks in different stages of alteration from volcanic heat, acid vapours, and atmospheric injury.

This singular combination, however, exists but on that portion of the puy which fronts the south-west, south, south-east, and east. The remainder is a massive and nearly vertical bed of domite alone. The whole mountain rises out of the semi-crater of the Puy de la Goutte, which closely embraces it from north-west to east, passing by the south.

4. Puy La Goutte from the East.   Puy Chopine.   Débris.

5.   1. La Goutte.   2. Chopine.   3. Chaumont, as seen from the South.

I spent some hours in examining the Puy Chopine at three different visits; and it appeared to me that, notwithstanding the confusion which reigns through some of its parts, the disposition of the principal masses is sufficiently obvious, and may give a clue to the ænigma of its formation.

The lowest rock which shows itself on the face of each rent above the talus of débris that conceals the base of the mountain, is a conglomerate of scoriæ and volcanic ashes, through which protrude massive blocks of basalt, evidently *in situ*. This basalt is either cellular and scoriaceous, or affects a spheroidal structure, contains olivine and augite, and puts on the dull aspect of the older rocks of this genus.

Upon this rests a tabular bed of granite. The line of contact is straight and inclined to the horizon at an angle of about 15 degrees, traversing the whole face of the mountain, from south-west to south-east. The granite for the space of three or four feet from this line is stained of a red colour, and so entirely disaggregated that the foot sinks into it; it becomes sounder as it is more distant from thence, but its masses present everywhere an irregular and dislocated appearance. Higher up it passes on various points—1, into a small-grained granite; 2, a sienitic rock similar to one met with near the Lac d'Aidat; 3, a fine-grained hornblende slate separating into rhomboidal pieces; 4, compact felspar.

These transitions are effected suddenly, and only remarkable here for occurring within such narrow limits, as the same passages on a larger scale present themselves on other points of the gneiss district. The assembled primary rocks are backed to the north, and in part supported to the west, by a rocky mass of domite which constitutes all that side of the mountain.

At the summit, and wherever this junction takes place, which is far from being as regular and decided as that with the basalt below, the granitic rocks are more or less altered. The granite is blanched, and has lost its consistence; the hornblende rock and compact felspar are also discoloured, cracked, and their fissures coated either by a reddish-brown ferruginous varnish, or dendrites of specular iron, evincing that the rock has been traversed and affected by volcanic exhalations. Small patches of boles and breccias, which are seen on a few of these points, seem merely to result from the mechanical confusion, decomposition, and chemical changes wrought in the united rocks already mentioned during their protrusion.

It appears then that the Puy Chopine, taken in general, consists of a mass of various primary granitoidal rocks, showing signs of great disturbance, included (like the meat in a sandwich) between a bed of domite on one side and of basalt on the other; and this singular aggregate rises immediately out of the crater of a volcanic cone of loose scoriæ.

These general features are indeed unparalleled. They leave, however, no room to doubt that the whole mass was raised into its present position by the eruption which threw up the scoriæ that form the Puy de la Goutte. Among the clefts which are broken across the superficial crust of rocks, by the expansive force of subterranean lava, and through which all volcanic eruptions take place, it must sometimes happen that two fissures meet or branch off at an angle, and occasion the minimum of resistance upon that point. An eruption taking place there may readily be supposed to elevate a portion of the rocks forming this angle in a solid state, and set it on edge upon one side of the vent. This mass would then form an obstacle to the fragments and scoriæ thrown out from the vent on that side, and force them to accumulate into a semicircular ridge on the opposite side of the orifice of eruption, such as is seen in the Puy de la Goutte. The basalt which underlies the primary crystalline rocks dates, no doubt, from this eruption, and is the outcropping of the subterranean lava whence the aëriform explosions proceeded. With respect to the trachyte, which also in part underlies the granite and backs it to the north and west, it is difficult to determine whether it owes its formation to the same eruption, or existed previously in contact with the granite, and was raised into its present position at the same time with that rock. I am inclined, however, to prefer the first opinion, and to think that we have perhaps in this close union of primary crystalline rocks, trachyte, and basalt, in a mass protruded from the crater of a volcano, an example of the *contemporaneous* elaboration, at a recent epoch, of the two most frequent varieties of volcanic rocks, trachyte and basalt, the first from highly feldspathic granite, the second from hornblende rock, by volcanic influence. If we suppose, which is not improbable, that the inferior and intensely heated beds of the original rock contained the same varieties of mineral composition as its superficial parts still exhibit, we may conceive how the same process, acting on the more feldspathose portions, would convert them into trachyte, which changed the more ferruginous or hornblendic parts into basalt; the quartz in each being dissolved and partly carried off by the aqueous vapour. These different lavas would intumesce, and might rise almost at the same time through the fissures on either side of the angular and still solid portion which was heaved up by the violence of their escaping efforts; and thus this portion of primary crystalline rocks would remain, as it now appears, wedged in between outcropping beds of trachyte on one side, and basalt on the other. Such a solution of the ænigma offered by the phenomena of the Puy Chopine, corresponds certainly with the phenomena of the mountain, and seems to be in harmony with the laws of volcanic action.

15–19.—North of the Puys de Chaumont and Chopine is a group of seven or eight volcanic cones scarcely separated from one another, and in all appearance the product of contemporaneous eruptions from different points of the same fissure.

The two which rise in the centre of this group, united by a narrow ridge, far exceed the others in height and magnitude. Each has a large crater at its summit; that of the most northerly, the Puy de Jume, is 210 feet in depth, and beautifully perfect; the other is called the Puy de la Coquille.

This system of volcanic mouths has sent forth a vast torrent of lava towards the east; which, entering a branch of the valley of Arniat, by this opening descended to the alluvial and calcareous plain below, on which it forms an extensive plateau, raised very considerably above the surface of the plain on either side; thus presenting an admirable clue to the formation of some more elevated and older basaltic plateaux in the vicinity, whose connection with the granite heights, whence they probably flowed, having been destroyed, might, but for this object of comparison, have appeared problematical.

The rivulets which drain the hills of Arniat and Laqui still unite in their former channel, and flow on unseen beneath the bed of lava, till at its extremity the whole river gushes forth near the villages of Sayat and St. Vincent, in the most abundant and fertilizing springs of the country. This basalt encloses numerous crystals of augite and olivine, with a few of felspar and some plates of mica.

20. *Puy de la Nugère.* — A volcanic cone, bearing traces of several craters: the principal one is an oblong basin, large and deep. It pours forth a body of lava towards the northwest.

Numerous smaller orifices seem to have been in action at the same time, and, after raising in common the north-eastern projection of the mountain, to have furnished another mighty torrent of lava, which is seen to fall in a broad cascade down its steep sides, and, encircling a prominent knoll of granite that checked and divided the current, to join the former stream.* Together they inundate a considerable valley, and, escaping

---

* See Plate IV.

from it through a narrow pass, descend to the site of the town of Volvic, where they are either stopped or lost beneath a later current from the neighbouring Puy de la Bannière.

The rock of Nugère has for three or four centuries past been quarried, and in use as the principal building-stone of the country, under the name of "Pierre de Volvic." It is throughout penetrated by numerous irregular cells *lengthened in the direction of the current*; it contains a large proportion of felspar, so much so as to make it approach in character to trachyte; in fact, is scarcely distinguishable from some of the trachytes of the Mont Dore, and bears a considerable resemblance to the Neidermennig mill-stone lava rock, near the Laacher See, in the Eiffel, as well as to some Italian lavas, especially that quarried near Camaldoli, of which extensive use is made in the town of Naples. It is of a light grey colour, and possesses the property of cutting with facility. The fissures occasioned in the mass by its retreat are few and distant. Hence blocks of great dimensions can be extracted. I have seen some worked into slabs of 20 feet by 6. These fissures are lined with a profusion of specular iron, which is also thickly disseminated through the pores of the rock.

21. *Puy de Louchadière.*—This is, next to the Puy de Dôme, the most striking hill of the whole chain. Completely isolated from the others, it rises as a majestic cone to the height of more than 1000 feet from the western plain, at an angle of 35°, and to the absolute elevation of 4000 feet.

An enormous crater, broken down towards the west, and measuring 486 feet in vertical depth from the highest point of the ridge, is scooped out of this mass. At its bottom a capping of basalt conceals the orifice of the volcano, and almost seems still to boil up from it as a spring gushes from its source: thence issues a vast current, which, first falling abruptly down a steep declivity into the plain, encumbers a wide extent of it with hilly waves of black and scorified rocks, and then, pursuing the direction of the slope, proceeds to join the more recent Cheire of Côme immediately above Pont Gibaud. The figure of this mountain originated its appellation: "La" or rather "Lou Chadière" is in Auvergnat "*the arm-chair.*" It is covered with forests, which add considerably to the beauty, and take from the horror, of its aspect.

22–25.—From the base of Louchadière a group of five or six smaller volcanic hills, closely connected, extends upon a line towards the north. They present the vestiges of as many craters, which send forth currents of lava to the east, west, and south. The thick forests which clothe these puys render it difficult to examine them minutely. Their lavas are uniformly basaltic.

26. *Puy de Pauniat.*—This cone stands by itself to the north-east of the latter group. It has a semi-crater facing the north-west, and seems to have furnished its contingent to the sea of lava which deluges the plain in that direction, but which later showers of volcanic ashes, a coating of rank grass, and here and there the plough, have combined to smooth over and conceal.

27. *Puy de Thiolet.*—An imposing hill of an oblong figure partly covered with wood. Its principal crater is elliptical and of vast dimensions. This cone has emitted streams of lava from three of its sides. That which issues from below the eastern base was anciently quarried, and the cathedral of Clermont was built from it; it was afterwards discovered that a similar stone was to be found nearer that town, viz. at Volvic, and these quarries were abandoned.

28. *Puy de Beauny.*—At some distance to the north-east of Thiolet is a circular plain about a quarter of a mile in diameter, once the bottom of a piece of water called the Lake of Beauny,

which a few years back was drained. On the south it is overlooked and partly embraced by a volcanic cone, which exhibits a crater breached towards the lake-basin. On the north and east it is enclosed by a semicircular ridge of loose scoriæ and granitic fragments, about 150 feet in height, on the outside of which is seen the granite.

This wide crater appears to have been opened through the granite by some more than ordinarily violent explosions, and the mouth was afterwards, in all probability, closed by the lava of the Puy de Beauny, which, first filling the basin, issues from it towards the west, and unites itself to the western streams of the Puy de Thiolet.

The lavas of the last described seven or eight puys, hemmed in by ranges of granite on the north and west, and by their own ejections on the east, seem to have flooded the whole space thus enclosed with their united currents. They resemble one another very closely in composition; and this similitude extends to those of La Nugère and Louchadière. They are of a light colour, abound in felspar, are more or less cellular, highly sonorous, and contain few or no imbedded crystals. It is probable, indeed, that these puys all date from the same epoch, and were produced by the same crisis of subterranean effervescence.

The connected chain of puys ends here; but two eruptions have taken place north of this point upon the same meridian, and probably therefore on a continuation of the same primary fissure of eruption.

29.—The first is the *Puy de Chalar*: it rises at a short distance east of the small town of Manzat; and three miles and a half in a direct line north from the Puy de Beauny. It is a large and sufficiently regular volcanic cone, with a vast crater broken down to the north-west. From its interior a copious current of very black, compact, and scorified basalt takes its rise, and descends by a gentle acclivity into the valley of the Morge.

30.*—Rather within a mile to the north-east of the Puy de Chalar is a circular lake, called *Le Gour de Tazana*, about half a mile in diameter, and from 30 to 40 feet deep. Its margin for a fourth of the circumference is flat, and little elevated above the valley into which the lake discharges itself. Everywhere else it is environed by a crescent of steep granitic rocks rising about 200 feet from the level of the water, and thickly sprinkled with small scoriæ and puzzolana. These fragments are all that indicate the volcanic origin of this gulf-like basin, but these are enough. No stream of lava or even blocks of any size are perceivable.

The encircling rocks show marks of considerable disturbance. They consist of two varieties of granite; one fine-grained, and of the usual elements; the other coarse, with very large crystals of felspar, and having the mica chiefly replaced by pinite, both crystallized in hexahedral prisms, and amorphous.

This curious, and, in Auvergne, rather uncommon variety of crater is identical in characters with some of the largest and most remarkable of the volcanic *maare* in the Eiffel (particularly that of Meerfeld); with this only difference, that the former has been drilled by the volcanic explosions through granite, the latter through superficial strata of grauwacke slate and secondary sandstone.†

The peculiar characters by which such craters are distinguished from the other volcanic vents we have been describing, viz. their great width, the total absence of all lava-current, and the ex-

---

* Both of these sites of eruption are without the area of our map of the Monts Dôme.

† See a paper on the Rhine Volcanos, in the Edinburgh Journal of Science, June, 1826.

tremely small quantity of ejected scoriæ by which they are surrounded, are not easily accounted for. If, however, we conceive an accumulation of highly elastic vapour to have formed on the surface of a subterranean reservoir of lava, like an immense bubble, and then by reason of an increase of its temperature to have exploded in perhaps a single or a few violent eructations (like the explosion of a steam boiler), the displaced superficial rocks falling back for the most part into the cavity, we should expect some such result as this kind of maar-crater. The extremely regular circular figure, which is characteristic of all such craters, demonstratively proves that the explosion of one or more elastic bubbles was concerned in their production. Had there been a lengthened series of such explosions, as in the usual phase of volcanic eruption, these would necessarily have thrown up a larger quantity of scoriæ and other fragmentary matters, and formed the usual circular ridge or cone round the crater; whereas in these instances such ejections are almost wholly wanting.

The Gour de Tazana is not quite a solitary instance of this phase of volcanic development in the chain of puys. The Lac de Beauny, already described, seems to owe its origin to some similar explosion on a minor scale; and we shall have occasion hereafter to remark its repetition in more than one locality of the Mont Dore.

Eastern Line of Puys.

Three other volcanic cones occur near the eastern limit of the granitic plateau, at some little distance from the meridian on which rises the chain of Puys hitherto described.

31. *Puy de la Bannière.*—An eruption has, in this instance, taken place on the summit of a granitic eminence overlooking the valley-plain of the Limagne, and a vast current of lava has

taken that direction, rushing tumultuously down the side of the mountain, and spreading over a large level surface below.

No regular cone or crater is to be observed; but prodigious heaps of scoriæ, volcanic bombs, lava-blocks, and fragments of granite, surround the spot whence the current issued. The shelving sides of the hill might, indeed, be expected to prevent the accumulation of these incoherent substances in the figure they usually affect. The town of Volvic is built upon the basalt of La Bannière, which crosses and apparently covers that of La Nugère. Copious springs gush from below it both at St. Genest and Marsat. It is of a black colour, compact, contains numerous crystals of augite, and fewer of olivine and felspar; and in all its characters contrasts with the lava of Nugère, bearing the strongest similitude to the basalt of some of the most ancient plateaux.

32. *Puy de Channat.*—This cone is partly covered with wood, which perhaps conceals the vestiges of its crater, nowhere else observable. Its scoriæ frequently envelop fragments of granite, and large knots of olivine and hornblende. The latter are rounded and amorphous, appearing as if they had suffered bouldering, or external friction. They are of a deep black, and exhibit a laminar structure only on fracture; the faces of the laminæ are highly lustrous. Crystals of the same mineral, with grains of compact felspar of a bluish-white colour, are to be found in the lava of which this puy has furnished two currents. One descends directly to the east, and is lost immediately under a cultivated soil; the other directs itself at first to the S.W., but is conducted by the inclination of the ground into the bed of a small stream, which it continues to occupy, making the circuit of a granitic hill covered by a fragment of a more ancient basaltic current, and disappearing towards the east below the village of l'Etang.

33. *Graveneire.*—This volcanic cone has been more frequently visited and described than any other, from its immediate vicinity

to Clermont. The eruption to which it owes its origin burst through a bed of basalt, which covered the eastern slope of a long crescent-shaped granitic eminence, called the Puy de Charade, one of the highest points of the plateau.

The scoriæ, lapilli, and puzzolana, of which the cone of Graveneire consists, have an exceedingly fresh appearance: they are red, reddish-brown, and black, and are often met with in the form of bombs, tear-drops, and long ropy sticks. The puzzolana is in great request as an ingredient in the mortar of all the neighbouring edifices: it is called "gravier noir" by the natives, and hence the mountain's name.

No crater is visible; probably it was destroyed during the emission of the last current of lava, which seems to have descended from the very summit of the present cone towards the north into the valley of Royat, stretching thence into the plain as far as Mont-joly and Les Roches.*

Two other streams are seen to proceed from the midst of scoriæ on the top or side of Graveneire, towards the east and south. The former is diverted from its course by a calcareous eminence, on which rises the Puy de Montaudoux, a conical rock of older basalt, and both unite to deluge an extensive surface of the plain below.

The interior of the bed of basalt in the valley of Royat is disclosed on each side of the vast excavation already effected through it by the rivulet. It measures 65 feet in thickness, and is divided by vertical fissures into imperfect prisms, or polyhedral

---

* See Plate I. Those geologists who, for the purpose of supporting the strange theory of "elevation craters," affect to deny that a stream of lava can harden on the slope of a hill at an angle above 4° or 5°, should examine those of Graveneire, which rest in beds at least 10 feet thick, at angles exceeding 40° on the upper part of the cone. And yet the bulk of this lava was so liquid as to flow to the distance of 4 or 5 miles, and spread in a wide sheet over the plain beneath.

blocks, strongly resembling in general appearance the forms assumed by many granites. These blocks are sometimes replaced by globular concretionary masses, with a concentric lamination, of which the woodcut annexed affords an example taken from a

6. Lava-Rock of Graveneire, showing an imperfect Prismatic and Spheroidal Structure.

rock near Royat. From some of the fissures of this lava-rock spring the copious sources which by means of an aqueduct supply Clermont with an abundance of the clearest water.

Within the gardens of Mont-joly, at the extremity of the northern current, is a small cavern rivalling the Grotta del Cane in its phenomena. A constant emanation of carbonic acid gas takes place from its sides and bottom; its mephitic qualities have been ascertained by repeated experiments; and in the village of Royat, at the northern verge of the lava-bed, very copious hot springs have lately been discovered, and applied to the purpose of an establishment of baths, built on the site of one formed there by the Romans, as attested by considerable remains.

The basalt of Graveneire, taken from the inner parts of the current, is compact, dense, of a deep slate colour or greyish-black, and contains crystals of augite, with a few of olivine and glassy felspar. The superficial parts are more or less cellular and scorified. Its texture is sometimes as highly crystalline as any of the oldest basalts.

From its proximity to the populous town of Clermont, the surface of the Cheire of Graveneire has been forced into cultivation by the most assiduous industry. The process is to break up all the projecting masses of basalt by blasting; and from their fragments and scoriæ, aided by dressings, a soil has been created and clothed with vineyards, which almost rivals the well-known fertility of the sides of Ætna and Vesuvius, where the same method has been constantly pursued.

Puys facing the Puy de Dôme.

34. *Puy de Colière.*—About 200 yards from the eastern base of the Puy de Dôme rises a diminutive cone of this name, which, seemingly from its insignificance, has as yet escaped the notice of every writer on the Monts Dôme. It has, however, to all appearance, sent forth a considerable stream of lava, which, after spreading to some extent over the plateau, encountered a hill of granite above Font de l'Arbre: a division took place, and one portion joined to the left the current of Pariou at Les Cheix, while the other threaded the narrow valley of Fontanat, and accompanied its rivulet to Royat, where it either terminated or has been since covered by the stream descending from Graveneire.

The basalt of the current of the Puy de Colière is dark-coloured, dense, brittle, and remarkable for being generally replete with numerous nearly spherical cells. Its crystalline texture is evident to the naked eye, and exhibits a multiplicity of minute crystals of glassy felspar; those of augite are not so conspicuous, but some crystals of this mineral and of olivine are scattered through the mass.

35. A short distance south of this is a still smaller cone, known to the shepherds by the name of Chuquet * Geneto, probably from the broom (*genêt*) which grows upon it. It has produced no lava, and was perhaps thrown up by some casual explosions finding a vent in this direction during the more violent eruption of the neighbouring puys. Its scoriæ, which are very fresh, contain many imbedded fragments of granite at different stages of alteration by heat, and nodules of hornblende similar to those of the Puy de Channat.

36, 37. *Puys du Petit et Grand Sault.*—Two small cones west of the Puy de Dôme, in which it is difficult to discover traces either of craters or lava, from the quantity of turf which clothes and surrounds them.

Westward of the river Sioule three or four other volcanic cones rise from the primary platform, viz. the Puys de Banson, de la Vial, and de Neufont. Not having visited these, I can give no details on their phenomena.†

Chain of Puys south of the Puy de Dôme.

38. *Puy de Gromanaux.*—An eruption has in this instance, as was noticed before, forced its way through a bed of domite, the prolongation of the base of the Puy de Dôme. A vast semicrater facing the N.W. discloses this rock *in situ*, and it projects also on other points of the puy.

---

* The term "Chuquet" acts as the diminutive of Puy, and is applied by the Auvergnats to any small knoll or cluster of rocks.

† They are mentioned cursorily by MM. Bouillet and Lecoq in their 'Vues, &c., du Département du Puy de Dôme,' p. 219, but I have met with no other account of them.

From the midst of this crater rises—

39. *The Puy de Besace.*—A double cone covered with turf, and, like the last and other neighbouring puys, presenting numerous fragments of domite mingled with its scoriæ and lapilli. It perhaps contributed to the basaltic lava-currents which have deluged the western slope of the granitic plateau.

40. *Puy de Salomon.*—A wide crescent, half encircling a crater which has pushed forth a large stream of lava to the west. Erratic fragments occur on this hill of a compact basalt, of light grey colour, and granitoidal texture so coarse as to exhibit its imperfect crystals to the naked eye. Three-fourths of these are of glassy felspar, the remainder of green augite, with a few of bright yellow olivine. The cellular cavities of this lava are lined with more perfect crystals of the same substances.

41. *Puy de Montchié.*—Four volcanic mouths combined to raise this mountain. Its northern crater is of vast diameter, and 340 feet in depth. That to the south-west appears to have given rise to the current of lava which descends below Alagnat.

The great proportion of domitic fragments which enter into the composition of this and the neighbouring hills, and which are scattered over the plain on each side, lead to the presumption that a considerable bed of this rock once existed on their line, and has been subsequently broken through, and launched into the air by their explosions. The road cut along the base of the Puy de Montchié has uncovered many charred trunks of trees buried in the dejections of the volcano.

42. *Puy de Barme.*—This volcanic hill stands by itself, a little to the west of the chain. It has three distinct craters; two at its summit which are perfect, and a third broken down towards the south-west. From hence issued an immense stream of lava, which widely flooded the western slope of the plateau,

and reached the Sioule on the site of Pont des Eaux. Usurping the channel of this river, it flowed on towards the north, and stopped beyond the village of Olby.

M. de Montlosier, who so well describes the changes wrought in the direction of the Sioule by the lava of Côme, seems not to have observed this third invasion, which has forced its waters to wear themselves a new bed through the hill forming their western bank, and consisting of a clayey alluvial tuff, a ramification of the Mont Dore.

43. *Puy de Laschamp.*—This hill when viewed from a distance has nothing of a volcanic appearance. It is a long-backed ridge formed by the united ejections of three or four craters, which have considerably defaced one another. Its highest summit, seemingly the result of the last eruption, is nearly 1000 feet above the village of Laschamp in the plain immediately below.

On the north exists the greater part of a large oblong crater; and a current of lava, springing from its base, turns to the east, and covers a considerable space enclosed between the Puy and the opposite heights of granite. Overflowing here, one portion of the current takes to the left, skirts the base of the Puy de Dôme, and descends to Enval, where it joins that which derives from the Puy de Colière already noticed; the other enters the little valley of Beaune, but disappears before it arrives at Fontfredde, beneath the meadows of the low ground it occupies. The basalt of this current is cellular and coarsely crystalline, of a light grey colour, very similar to that of the Puy de Salomon above described, has a large dose of felspar in the composition of its base, and imbedded crystals of this mineral and of olivine. A separation into rude columns is observable on some points. Another semi-crater to the left has also emitted a stream of lava, which, after joining those of the Puys Lamoreno and Montchar, two dependencies of Laschamp closely united to it, spreads over a part of the western slope.

To the south of the Puy de Laschamp rises an irregularly circular system of volcanic cones, the produce of many repeated eruptions within a small space, which in all probability succeeded one another very closely, or raged at the same epoch. This is the most interesting portion of the whole range to every observer, whether geologist or not. The extraordinary character of the view from any one of these puys impresses it for ever on the memory. There is no spot amongst the Phlegræan fields of Italy or Sicily which displays in greater perfection the peculiar features of a country desolated by volcanic phenomena.

It is true that the cones thrown up around are partially wooded and in general covered with herbage; but the sides of some are still naked, and the interior of their broken craters, rugged, black, and scorified, as well as the rocky floods of lava with which they have loaded the plain, have a freshness of aspect, such as the products of fire alone could have preserved so long, and offer a striking picture of the operations of this element in all its most terrible energy.*

The first cone which presents itself commencing on the east is,

44. *The Puy de Mercœur.*—It has a small crater on the summit, and is not otherwise remarkable.

45. *The Puy Noir* or *de la Meye,* which occurs next, has a large semi-crater 600 feet deep facing the east; from its interior proceed one, or rather two, of the most bulky lava-currents of the Monts Dôme. After covering a wide space above Fontfredde with a sea of feldspathic lava, this upper current stops there, and a lower bed of black augitic lava shows itself from beneath, enters the valley of Theix, and occupies the winding bed of

---

* See Plate V.

its rivulet as far as Julliat, where it terminates a course of ten miles, performed with a fall of 1700 feet. Copious springs gush out at Fontfredde from beneath the upper bed of lava.

The scoriæ of the Puy Noir are exceeding dark coloured, and justify its appellation. The lower lava contains frequent crystals of augite and olivine; and where the current has been obstructed by rocks of granite narrowing the gorge through which it flowed, an imperfect columnar configuration may be observed.

46. *Puy de Las Solas* or *de la Gravouse*.—This cone encloses a widely breached crater. Its steep walls of black and crumbling scoriæ half surround an abyss, which almost seems still to vomit forth an impetuous torrent of lava. This immediately joins that of the

47. *Puy de la Vache*,—a ruined crater, the exact counterpart of the last puy, to which it is united by the base.

The whole appearance of these two remarkable hills forcibly demonstrate that the showers of scoriæ which created them were thrown up previously to the emission of any lava; and that this substance, afterwards rising in a state of liquefaction through the chimney of the cone, filled the funnel-shaped cavity of the crater, broke down by its weight the weakest side, and rushed forth to deluge the surrounding soil. In the interior of the upper part of the remaining circumference of the Puy de la Vache, the point to which the lava rose in the crater is still marked by a projecting ridge of light scoriaceous matter of a reddish yellow colour, rich in specular iron, and considerably decomposed by sulphureous vapours; apparently part of the frothy scum which formed upon the surface of the ebullient lava, and adhered to the side of the vase at the moment of its being emptied.

A stopper of basalt still chokes up the orifice below, and connects itself with a vast current, which, being swelled by the addition of those of Las Solas and the Puy de Vichatel immediately opposite, takes its course towards the south-east, and by damming up the channel of two rivulets near their confluence has given birth to the two lakes of La Caissière and d'Aidat, the latter of which is a large and picturesque expanse of water.* From this point the stream of lava entered and threaded a narrow granitic gorge, and, swelling out again as the valley widens towards St. Saturnin, filled its basin completely, and stopped where Tallende is now situated. The distance traversed by the lava is about twelve miles, and the difference of level between its source and termination 2230 feet. Near the spot where the town of St. Saturnin stands, a body of stagnant water appears to have existed at the period of this irruption; for beneath the bed of basalt, there about forty feet deep, a stratum of clay shows itself, containing an abundance of vegetable remains reduced to charcoal by the intense heat of the superincumbent lava. It is remarkable that the clay immediately in contact with the basalt is hardened, and divided to the depth of ten or twelve inches into small vertical prisms, exactly imitating in miniature the columnar groups of basaltic plateaux.†

The Cheire of La Vache, &c., is particularly ragged and bristling, as indeed are those of all the puys of the southern extremity of the chain,—a character they owe to the prevalence of augite in their composition; while the lavas of the northern

---

* Sidonius Apollinaris is said to have had his habitation on its banks.

† It is clear that the retreat which in the one is effected by the loss of caloric, is in the other caused by its increase. The principle of fluidity in either case disappears, and a contraction must take place.

I observed the same sort of natural equivoque in frequent instances in the Velay; where the odd conjunction of these different substances, in which the same effect is produced by directly opposite causes, was rendered more striking by the equally regular columnar structure of the clay, and the basalt which had hardened it.

puys, as we have seen, are almost wholly composed of felspar, and consequently present much fewer asperities.*

Some of the scoriæ of the Puy de la Vache appear to have suffered an alteration from the action of acid or sulphureous vapours; they are found of various tints of white, yellow, blue, red, and black; they abound in specular iron, which is also found to have insinuated itself by sublimation into the fissures of compacter lava-blocks, forming within them the most delicate dendritic ramifications.

48. *Puy de Vichatel.*—This cone has a regular crater sloping towards the north-east. Its lava mingles with the currents of La Vache and Las Solas.

49. *Montchal.*—A regular cone with a semi-crater fronting the north; it is partly wooded. Its lava probably is concealed by that of

50. *Montgy.*—This puy has sent forth a considerable stream of lava from a crater broken down towards the south-east, exactly in the direction of Montchal; which appears to have divided the stream into two branches. They are lost immediately under the ashes and other loose matters with which the plain around has been strewed by subsequent explosions, and the vegetable soil to which their decomposition has given rise.

51. *Pourcharet.*—A large and imposing cone, the most westerly of the group. Its sides are steep; and on the summit is a wide but shallow crater. Lava has issued from its north-western base, and followed the slope of the plateau to some distance in that direction.

52. *Montillet.*—A crescent-shaped ridge of little elevation; from the semi-crater it encloses proceeds a current of lava, which turning to the right joins that of Pourcharet.

---

* See p. 65 above.

53. *Montjughat.*—This puy stands by itself, and is one of the most regular of the chain. The crater is deep and large. Its lava appears to have joined that of Montillet.

54. *Puy de la Taupe*—so named from the resemblance which it bears, in common with many of the rest, to the labours of some giant mole. It has a crater breached towards the west, and its lava covers a considerable surface on that side. The basalt of this current is remarkable for containing even more crystals of augite than that of the neighbouring puys. It is exceedingly dark-coloured and compact.

55, 56. *Puys de Broussou* and *de Combegrasse.*—These cones, closely united, have craters open to the south-east, and furnish a current which turns to the east and reaches that of La Taupe.

57, 58. *Puys de la Rodde* and *de Chalard.*—The former is a large hill apparently produced by more than one volcanic mouth. It has a semi-crater facing the south, and gives birth to a current of lava which spreads to the right and left, reaching the village of Aidat, at one extremity of its lake.

The scoriæ of the Puy de la Rodde are remarkably rich in crystals of augite of the most perfect regularity, and the basalt of its current equally abounds in crystals of this mineral and of olivine.

The Puy de Chalard appears to have been thrown up by a lateral aperture, immediately after the eruption which created the Puy de la Rodde.

59. *Puy de Charmont.*—This cone has a large and deep crater broken down towards the south-east; from the bottom issued a current of dark-coloured lava, but was prevented by a ridge of granite from reaching the lake of Aidat.

60. *Puy de l'Infau* or *l'Enfer.*—A low but large crescent, the remaining walls of a vast crater, whose site is now occupied by a circular bog called La Narse d'Espinasse, from the nearest village.

The plain to the east and south is formed of ancient beds of basalt, and a violently explosive eruption has here evidently made its way through the continuation of these: their sections may be observed round the interior of the crater. A current of lava has been vomited by this opening towards the east.

61. *Puy de Montenard.*—The most southerly of the chain of puys belonging to the Monts Dôme appearing in our map. Its figure is irregular, the result of two neighbouring craters. One of them is open to the south-east, and a copious stream of lava has issued from it, covering a wide space on the south and east, and reaching to the bed of the river at Monne. But the cultivation of the plain from which this puy rises, and the quantity of more ancient lavas which have flowed from the Mont Dore in its direction, renders it difficult to ascertain the boundaries of the recent currents with due precision. Although the continuous chain of puys ends here, another very recent cone and lava current, called Tartaret, occurs on the continuation of the same meridian at the distance of about three miles. But as that point is within the circuit of the system of Mont Dore, its description is postponed for the present.

VOLCANIC ROCKS THE PRODUCT OF EARLIER ERUPTIONS.

The eruptions that produced the chain of puys just described, though far from being all contemporaneous, evidently belong to a period during which the volcanic energy raged with peculiar fury along the line they occupy; a fury by which it appears to have been completely exhausted, and which may be the cause of its subsequent inertness.*

---

* Eruptions which occurred within the six years from 1730 to 1736 in the island of Lancerote, one of the Canaries, produced a chain of cones, some thirty in number, and a sea of lava, which deluged a vast surface. These craters

But a close observation convincingly demonstrates that previous to this era of intense activity eruptions were continually taking place within the same zone, without any quiescent interval of sufficient duration to enable us to mark out a decided line of separation between the recent and the ancient volcanic remains of the class that now occupies our attention.

For the convenience of description, however, I have thought it necessary to draw an arbitrary division between those already described as unquestionably *very recent,* and those to which we now proceed, which are of *earlier* production.

The former have been mentioned in the order of their geographical occurrence; the character and position of the latter will be best understood if we consider them in the inverse of what appears to be the chronological order of their formation, commencing with the products of the eruptions which seem to have immediately preceded those already described in the chain of puys.

They consist of currents of basalt, of which the originating cones and craters either no longer exist, or are partly obliterated; which are more or less denuded of their accompanying scoriæ or scorified surfaces; affected by decomposition; often divided and parcelled out into isolated plateaux or peaks; or finally, which overlook valleys or ravines of considerable depth evidently excavated since their formation.

67. The Puy Rouge near *Chalucet,* on the left bank of the Sioule, about two miles below Pont Gibaud, is one of the first connecting links between the recent and more ancient eruptions, possessing the principal characteristics of either class. It has

---

opened successively along a line stretching nearly the length of the island; no doubt the direction of a great subterranean fissure. They are described fully by Von Buch and Sir C. Lyell (Principles, 1853, p. 436), and present a remarkable parallel to the chain of recent volcanic cones of Auvergne.

burst forth from gneiss which environs it on all sides; and a considerable cone, composed throughout of lapilli, puzzolana, volcanic bombs, and scoriæ of every fantastic form, still marks the site of its orifice. At the northern base of this hill facing the Sioule, commences an enormous current of basalt, which evidently once flowed into and entirely choked the valley, extending more than two miles down the stream.

The river has subsequently worn itself a new channel between this accumulation of basalt and its northern bank; and the vast dimensions of this excavation eaten out of solid primary strata to the depth, in parts, of 50 feet below the level of its former bed, upon which the basalt is seen to rest, evince the long continuance of the erosive action, as well as its irresistible power. This amount of excavation can only be attributed to the *river* which still flows there, because the undisturbed and perfect state of the cone of loose scoriæ demonstrates that no denuding wave, deluge, or *extraordinary* body of water has passed over this spot since the eruption.

The massive cliffs of basalt that beetle over the gorge thus formed have a striking aspect. The thickness of the bed averages, perhaps, 150 feet; its upper surface is level, but covered with scorified protuberances, and, where it has not been forcibly brought into culture, as ragged as any Cheire of the Monts Dôme. These portions of cellular and shaggy basalt are, however, merely superficial; the interior and great mass of the current is compact, and remarkably divided into very small prisms grouped together in curved and radiated bundles. Beneath it is a bed of sand, gravel, and bouldered stones 3 feet thick, evidently the ancient bed of the river at the time of its invasion by the lava-stream, but now from 20 to 50 feet above the present channel, which the Sioule has since cut for itself through the gneiss beneath. In the latter rock several galleries have been opened below the lava, for the extraction of lead ore, which occurs in veins of the gneiss. Some of these were worked at a very early period, probably by the Romans; much ore is still extracted by a company recently formed.

On some points of this current the prismatic is replaced by a spheroidal concretionary structure on a small scale, the basalt separating at a touch into minute angular globules, rarely exceeding the size of a pea, the "pièces séparées grenues" of the French geologists. There is not the least probability of this structure having been produced or even disclosed here by decomposition; on the contrary, large horizontal masses of this nature alternate with others that are prismatic on a proportionately small scale; the latter modification appearing to pass into the former by the sole diminution of the axes of condensation. Where the basalt is seen in immediate contact with the supporting gneiss, this last rock has suffered a partial disintegration, and is stained of a bright red colour to the depth of a few inches.

In its mineralogical characters the basalt of Chalucet resembles some of the most ancient varieties; it is compact, heavy, of a dull aspect, a dark colour, and without any apparent imbedded crystals. It frequently envelops fragments of granite and mica-slate, considerably altered; and I have found amongst its scoriæ blocks of a porphyry similar to that of Chateix near Clermont.

68. *Puy de Charade.*—A considerable granitic eminence on the western edge of the plateau. Its summit is covered by a massive bed of basalt, prolonging itself with a rapid slope towards the east, till it is interrupted and concealed by the more recent volcanic cone of Graveneire, which appears to have exploded from beneath it.

On the opposite side of this hill, and at a lower elevation, the same bed of basalt reappears, and is easily recognised by its peculiar characters: it immediately encounters the still more ancient basaltic rock called the Puy de Montaudou,* and is separated by it into two branches, which reunite below, and follow the slope of the hill as far as the level of the plain.

A few scoriæ are observable on the summit of the Puy de Charade, sufficing to mark it as the site of an eruption, but nothing resembling a cone or crater. The basalt contains very large crystals of augite and nodules of olivine, has a dull and leaden aspect, is considerably decomposed, and assumes a frequent division into spheroids of a foot or more in diameter, which desquamate in concentric laminæ; yet, notwithstanding these numerous characters of great antiquity, its disposition in the form of a current of lava descending from the granitic heights into the main valley below is so evident, that M. Ramond felt obliged to rank Charade in the class of modern volcanos.†

The granitic sides of the mountain must, however, have wasted prodigiously, and the deep and wide ravines on either side been altogether formed, since the deposition of this lava-bed, which now hangs over them in a rocky ledge; while their sloping sides show them to have been excavated gradually by running water. Moreover, the main valley into which the lava

---

* See Plate I. The basalt of the Puy de Montaudou is very compact, fine-grained, hard, and black. It occasionally contains small granular crystals of felspar resembling those of Egyptian porphyry, and grains of peridot. The puy is a mere conical rock, of which it is difficult to ascertain the structure, from the vegetation which clothes it. Some rude prisms show themselves at the summit; and to the west it may be seen to rest on the freshwater limestone, the strata of which are much confused and tilted up at the line of junction. I presume it to be a dyke of basalt protruded on the spot from beneath the tertiary strata close to their junction with the granite. It is remarkable that this protuberance has evidently divaricated both the old lava-current of Charade and the more recent one of Graveneire in their descent from the western heights behind it.

† Nivellement des Plaines, 1815.

of Charade descended, almost to the level of the present plain, must have existed previous to the excavation of these minor valleys, and consequently it is impracticable to assume the formation of the valleys of this district in general as constituting an epoch or as marking any fixed period. Further proofs of this fact are to be met with on many points, as well of Auvergne as of the Velay and Vivarais. Their evidence will be discussed as we go on.

69. *Puy de la Roulade.*—A bed of basalt remarkable for its uniform separation into very complete spheroids, surmounting a small calcareous hill a few yards above the rivulet of Boiseghoux; and nearly on a level with the plateau formed by the more recent lava of Graveneire on the opposite bank. This might be supposed part of a secondary current from Charade, but that in mineralogical characters it differs materially from the basalt of that mountain. Its little elevation above the plain demonstrates the comparatively recent date of its formation.

70. *Plateau of Château-gay.*—An isolated bed of basalt covering a wide extent of the calcareous freshwater formation. Towards the Limagne it rises from the plain to an average height of 450 feet; but on the south-west the plateau formed by the lava of Jumes (16) equals it in elevation, and imitates it entirely in disposition and structure. The productive cause of the one is so evidently that of the other, that it is impossible to escape the conviction of their identity of origin. It is equally clear that the more complete insulation and greater elevation of the former above the plain is owing only to its having been longer exposed to meteoric erosion.

This current appears not to have had its source on the granite, but at a short distance from the limit of the two formations; and the site of its crater is still attested by the tumefied and cellular nature of the basalt of the north-west extremity, by the scorified masses, and, above all, by the numerous lava-drops or volcanic *bombs* (an infallible sign of an eruptive vent), which abound there. This is also the most elevated point of the plateau, and from hence it slopes gradually towards the south and east. It is generally modified into rude columnar prisms, which in parts evince by decomposition a spheroidal concretionary structure, and in others divide either into massive tables or slaty laminæ, between which arragonite is often found crystallized.

71. *Plateau of La Serre.**—This remarkable sheet of basalt owes its chief interest to the circumstance of its remaining almost entire, and exhibiting in consequence that peculiar disposition which would be assumed by a stream of fluid matter descending an inclined plane, and occupying the lowest levels, although it now forms the capping of a high hill-range.

That it flowed as a current of lava is moreover attested by the scoriæ upon which the bed of basalt rests, by the scorified and cavernous masses which still remain on its surface, but which adhere to and form part of the compact and solid rock below, and by the site of the vent which produced it being still cognisable on its highest extremity.

On the other hand, this basaltic bed has all the characteristic features of the more isolated and hitherto contested plateaux.†

---

* See Plate I.

† It must be remembered that at the date of my examination of this country (1821-5) the contest was still raging between Neptunians and Vulcanists as to the igneous or aqueous origin of the flœtz traps, and that even those who admitted their igneous origin still insisted that all the flat sheets of basalt so frequently found capping high plateau-shaped hills were formed under the sea, and in totally different circumstances from those characterising a recent volcanic sub-aërial eruption. I have retained in this edition some passages like that above, in which these doctrines, now all but exploded, are controverted, because even yet the fallacy, I believe, is scarcely extinct that horizontal and extensive sheets of basalt are almost necessarily of subaqueous origin.

Its elevation varies from 850 to 400 feet above the water-channels of the valleys on either side : and a branch of it, cut off and separated from the main current by a subsequent excavation, has assumed that conical form so general amongst basaltic remains, and to which the waste of ages tends to reduce all, and crowned by a ruined fortress, called Montredon, imitates exactly the Stolpens and Kœnigsteins of other basaltic districts.*

The current of La Serre originated in the granite, and the most considerable moiety of its extent rests upon that rock, the remainder on the freshwater limestone. Its western summit, called the " Tête de la Serre," or Puy de Nadailhat, measures 3461 feet from the sea-level. It terminates to the east in a projecting tongue of hill, the point of which has suffered a partial separation from the rest. On this point stands the village of Le Crest, at an absolute elevation of 2044 feet, giving a difference of level between the two extremities of 1417 feet, with a direct distance of rather more than six miles and a

---

* This process has been repeated on numberless points of Auvergne, where almost at every step we meet with isolated and conical peaks, each consisting of an immense group of basaltic columns converging towards the summit. All of these were seized on in turn for the sites of fortresses, in those times of anarchy when an inaccessible position was the necessary condition of security to person or property. Such were the castles of Mont-rognon, Montredon, Mont-rodeix, Mont-celets, Vodable, Usson, Nonnette, Buron, Mozun, Murol, Vandeix, Bonnevie, Mercœur, Ibois, Mercurol, &c. &c. . From the number of these strongholds and the almost impregnable nature of the greater part, the feudal tyrants of Auvergne outlasted those of the rest of France; and it was not until the ministry of Richelieu, and the vigorous reign of Louis XIV., that a final check was given to their career of violence and rapine. Many of them were judicially condemned and executed at Clermont by a special court held there and called "Les Grands Jours," a fate they well merited. Orders were then issued for the demolition of all the châteaux-forts of Auvergne, and little now remains of them but the foundations, and some fragments of their massive walls, which were generally constructed of basaltic prisms taken from the peak itself, and laid horizontally. Puzzolana was mixed with the mortar used in these constructions; and without the binding quality communicated by this ingredient, probably no cement would have taken effect on the smooth and iron surfaces of the prisms.

quarter. The inclination therefore of this bed of basalt corresponds almost completely with that of the two very recent lava-currents which have flowed down and now occupy the bottoms of the valleys on either side of it; the one proceeding from the Puy Noir, the other from the group of cones about the Lake Aidat :* moreover, the distance to which it reaches is about the same as in their cases. The parallel between the older and newer basaltic currents visible in such close approximation is complete and highly interesting. The only essential difference is that of position; the one occupying the summit of a long hill, the others the bottom of its lateral valleys, but yet on many points even there forming plateaux which have acquired already a very considerable elevation above the actual river-channels.

The surface of the plateau of La Serre is not a uniform slope, being broken by three declivities which have the appearance of steps. Two of these are over the granite, and another at the line of contact of this and the freshwater formation. They were probably occasioned by inequalities in the granitic surface here and there opposing a temporary check to the current. The portion which rests on the freshwater strata is smooth and nearly level.

It is worthy of remark that on each side a transverse ravine is found near where the tertiary sedimental beds rest on the primary crystallines, and a depression exists on the surface of the plateau, along that line, the basalt having been already partially undermined there by an excavation sapped through the strata of friable sandstone which intervene between the calcareous formation and the granite. In time, no doubt, an entire separation will thus be effected, and the eastern portion of this hill will resemble the many others around, which have lost their connection with the primary heights.

---

* See p. 92-3 *suprà*.

The basalt of this plateau is from 50 to 100 feet in thickness. Though in general amorphous, or cleft by irregular vertical seams, yet on some points, particularly near Le Crest and at the Castle of Montredon, it exhibits very beautiful columnar groups. The current has here been eaten into by deep ravines which disclose its internal structure, and in all probability the apparent absence of this regular configuration in the remaining parts of the plateau is only superficial.

Upon the whole, this hill is very instructive, as a type of the formation of basaltic plateaux in general; one of those valuable links which establish a relation between rocks apparently remote in geological position; one of the intermediate gradations through which a current of lava, on its first production, occupying the bottom of a valley, passes in the lapse of ages, and in virtue of the great resistance it offers (and lends umbrella-like to the strata beneath) to the downward wash of rain, into the massive and tabular capping of an isolated and lofty mountain.

72-73. *Les Côtes de Clermont and Chanturgue.*—These calcareous hills, once evidently united, and separated now but by a shallow ravine, are crowned with a bed of basalt, in nature and position very similar to that of Châteaugay; but its height above the plain is greater, and it has suffered more from waste, apparently in consequence of its superior antiquity.

It has a gradual slope from the neighbourhood of the granitic escarpment towards the centre of the Limagne, and appears to have flowed as a lava-current from the heights which rise on the north-west of Durtol: perhaps it may have been once connected with those remnants of basalt which rest upon the granite in the vicinity of the Puy Channat, but which have lost all traces of scoriæ or cellular parts.

74. *Plateau de Prudelle.*—A mass of basalt, which crowns a granite promontory impending over the valleys of Villar on the

29 AP 58

VALLEY OF AVELLAR AND PLATEAU OF TRIHUTE

▼ Town of Cintra.

Plate VI.

south and Gressinier on the north, and terminating in an abrupt escarpment towards the east. It appears, however, to have once extended some way down the steep slope in this direction, for many portions of a similar basalt are to be found *in situ* amongst the vineyards which clothe its declivity.

The cone whence this current flowed still forms a projecting knoll at its western extremity, strewed with bomb-shaped scoriæ, the cavities of which contain delicate stalagmitic concretions of quartz, the fiorite of the French mineralogists. It is this hillock which occasioned the division of the current of Pariou (6).

At the point of contact of the granite and the basaltic bed it supports, is seen a layer of scoriæ, so far decomposed as to be cut with a knife; their cavities are filled with a white and brown bole of a waxy consistence. The basalt has separated on some points into very regular prisms of five or six sides, which exfoliate by decomposition in slaty laminæ at right angles to their axes.

The accompanying engraving gives a view of the position of this plateau (on the left) hanging over the steep granitic gorge of Villar. Beneath is a Roman road, of which the old basaltic pavement is still entire (called le Chemin Ferré). It is formed upon the surface of the comparatively recent lava-current descending from Pariou, since the flowing of which, however, the ravine on the right has been worn away more than 150 feet in depth. It is evident that the entire gorge has been excavated since the basalt of Prudelle flowed upon the surface it now covers, which must then necessarily have been the lowest level of the vicinity. Here again is a most instructive collocation of the older and recent lava-streams, telling the same tale of the gradual erosion of the valleys of the district by causes still in operation.

Though the current of Prudelle has every appearance of being single, it consists of two very different species of basalt. That of the western extremity is remarkable for the numerous and large nodules of yellow and red olivine it encloses. That to the east has in place of this mineral confused crystals of augite and compact felspar, disseminated in a base which has all the characters of a fine-grained dolerite.

75. *Puy Girou* (A. E. 2888.). 76. *Puy de Jussat.* 77. *Gergovia.*—I consider these three eminences, all based on the freshwater limestone, to have originally formed a single plateau, capped by a current of basaltic lava proceeding from the neighbouring primitive heights, probably from the Puy de Berzé (79). (See Plate I.)

They are now partially separated from each other and from the granite; the first has wasted to a conical cluster of prisms converging towards the apex; * the second to a crested ridge; while the broad plateau of Gergovia still occupies an extended surface, which tradition and history unite to fix on as the site of that city of the Arverni which Cæsar and his legions so long and vainly besieged.†

Gergovia is as interesting to the geologist as to the antiquary. On the southern and western flanks thick strata of siliceous indusial limestone abound. The eastern face offers a distinct example of the alternation of basaltic currents with the calca-

---

* This peculiar disposition of the prisms, so evidently adapted to protect the mass they constitute from the destructive agency of rains and frosts, will be found on observation to be the cause which has preserved throughout all basaltic regions so many isolated and conical hills of this rock. I do not remember to have met with one of this nature in which it was not easy to trace such a structure; and this fact unites itself to the many others presented to the eye and understanding of the geologist by every great mass of mountains, which tend to force upon him an overwhelming conviction of *the vast amount of denudation of the supra-marine portions of the surface of the earth, that has been effected by the wasting powers of the meteoric agents,* since its emergence from the ocean. The great height of the calcareous strata at the Puy Girou (2800 feet, A. E.) is remarkable.

† Roman bricks, amphoræ, medals, and Gaulish axes and arrow-heads, in jade and serpentine, are frequently found on this plain.

CHAP. V.   ROCKS FROM EARLIER ERUPTIONS.   107

reous strata of the freshwater formation. Two deep ravines have there laid open complete sections of all its beds. The base consists, like the plain from which it rises, of thin horizontal strata of white marly limestone. At about two-thirds of the whole

7. Eastern Face of Gergovia, showing its two Beds of Basalt (1, 2), and the stratified Calcareous Peperino between them.

elevation occurs a massive horizontal bed of basalt, in some parts 40 feet in thickness, which appears to have moulded itself on the strata below, and upon the surface of which other calcareous strata have been again deposited.

These, however, though distinctly stratified, and with a general tendency to horizontality, are far from being as regularly disposed as the inferior limestone. They are frequently distorted and confused, and occasionally interrupted by narrow horizontal venous masses of basalt, which appear to be ramifications from the upper bed of great thickness, which surmounts the whole and forms the superficies of the plateau. The calcareous strata thus included between the two beds of basalt are thickly interspersed with volcanic ashes and scoriæ. A few thin strata of compact yellow limestone may be found apparently free from extraneous substances, but the general mass is rather a calcareous peperino than anything else. It is in parts veined with semiopal, and contains masses of siliceous limestone.

The lower bed of basalt projects far beyond the upper, appearing to have been originally more extensive; and those portions which cap the calcareous hills above Aubières, Perignat, and Channonat, must have belonged to this bed, which, whether erupted on the spot or at a distance, is clearly prior in date to the strata of calcareous peperino above it, as was remarked in an earlier page, and was produced at a time when the freshwater lake was still depositing its chalky sediment. It is in some parts very regularly prismatic, in others amorphous; is of a dark colour, dense, hard, and sonorous; contains minute scaly crystals of glassy felspar, and some calcareous infiltrations.

The line of contact of this bed of basalt and the supporting stratum of marly limestone is well defined, and specimens of the smallest size may be taken from it, of which one half is compact, brittle, and black basalt, the other a white limestone effervescing strongly with acids, and apparently uninjured. The basaltic masses which occur in the intermediate calcareous strata are far from being so distinctly separated from the enveloping matter. The one substance, on the contrary, seems almost to pass into the other by a mechanical intermixture effected when both were of a very soft, if not fluid, consistence.

The upper plateau of basalt rarely presents a prismatic division on a large scale; but more frequently a tabular structure. It is on some points extremely cellular; the vesicles having been subsequently filled by crystallizations of carbonate of lime and arragonite, so as to give it the character of a very rich amygdaloid.

78. *Montrognon.*—A conical eminence of columnar basalt, crowned by the ruins of a feudal fortress, and resting on the freshwater limestone. It is probably the sole remnant of a plateau formed by a branch from the current of Gergovia.

79. *Puy de Berzé.*—A salient eminence of the granitic platform; the site of an ancient volcanic aperture, for its summit is

thickly strewed with cellular basalt, scoriæ, and "bombs;" which, however, have an aspect of great antiquity. It was, in all probability, from this mouth that the currents of Gergovia, Montrognon, &c., proceeded. Their comparative elevation favours this opinion. The Puy de Berzé measures 3210 feet above the sea; the Puy Girou, 2792; and the western summit of Gergovia, 2496. The eastern extremity of the last hill is much lower, the whole surface sloping gradually in the supposed direction of its flow, i. e. from the granitic shore of the lake towards its centre.

80. *Basalt of St. Genest de Champanelle.*—This current occupies the bottom of a depression in the primitive plateau through which runs a small rivulet from the village of Chatrat to that of St. Genest. In its position, consequently, it agrees with the lavas of the recent puys; but since no cone remains to mark its source, and it has evidently suffered much degradation, it must certainly be prior in date to all these. It is imperfectly columnar, and in parts separates into minute angular globules; but the circumstance most remarkable in the basalt of this current is the large proportion of *quartz* which enters as an ingredient into its composition. This substance is very partially distributed, occurring in much greater abundance on some parts of the current than on others, and presents itself in three modes: 1, as visible grains, or imperfect crystals, imbedded in the base; 2, as a constituent part of that base, which observed through a weak lens appears to be a granitoidal mixture of quartz, felspar, and augite; 3, in distinct and frequently large veins, more or less free from augite, sometimes entirely pure, of a greyish-white colour, and similar in disposition and aspect to the veins of quartz so common in Lydian stone, clay-slate, &c. This is the only example of a quartzose basalt I am acquainted with in Auvergne.

81. *Puy de Chatrat.* 82. *Puy de Pasredon.*—Granitic emi-

nences capped by basalt, probably remnants of the same current, but certainly distinct from, and anterior to, that last mentioned.

83. *Puy de St. Sandoux.* (A. E. 2822 feet.) ("Barnère" of Ramond.)—A basaltic plateau, still more elevated but less extensive than that of Gergovia. It might perhaps be ranked among the dependencies of the Mont Dore; but as it forms a very conspicuous object in all views of the Monts Dôme, and enters into our map of this range, its description may more appropriately find room here.

It occurs on the limits of the granite and freshwater formation, resting in part on each. More than one current appears to have contributed to its formation. That which shows itself on the surface of the plateau is remarkable for presenting a complete transition from a coarse-grained dolerite, composed of compact felspar and large crystals of augite, to a perfectly homogeneous basalt. This change is effected both suddenly, so as to enter into the compass of a small specimen, and gradually, the crystals of augite and patches of compact felspar diminishing in size, and the latter being finally superseded by olivine.

84. *Plateau of St. Saturnin.*—A basaltic bed entirely on the calcareous soil, and at a lower level than that of St. Sandoux; but yet, perhaps, a branch of the same current. On the southeast, immediately above the Château of St. Sandoux, a portion of this bed has assumed the figure of an enormous spheroid, composed of columns diverging from the centre of the rock, near which they are closely united, to the circumference, where a considerable space is left between them. The columns are very regular and jointed. On other points the basalt of the same bed exhibits a tabular configuration.

85. *Puy d'Olloix.* (A. E. 3343 feet.)—A conical eminence of basalt, which apparently is a fragment of the same cur-

rent as the Puy de St. Sandoux. It encloses large knots of olivine.

The great comparative elevation of these last three basaltic eminences is remarkable. M. Raulin* considers that an elevatory movement on a line transverse to the general axis of the neighbouring mountain ranges has raised them, as well as the granitic and tertiary beds on which they rest, since their formation. This notion will be discussed in a later page.

86. *Chox de Coran.*—An extensive plateau of basalt resting entirely on the freshwater strata, and overlooking the Allier from its eastern extremity. On this side, immediately above the village of Coran, rises a vast range of basaltic columns, the upper portions of which show a tendency to the spheroidal figure. The south-western part of the plateau is covered with scoriæ and volcanic bombs, exceedingly fresh, and apparently ejected by a later eruption which burst through the more ancient basaltic bed. A circular cavity, about 30 feet in diameter and 10 or 12 in depth, appears to be the only crater left by this recent explosion. Its inner walls are perpendicular, and consist of scoriform masses of basalt, uncovered by any accumulation of lapilli or puzzolana; an inconsiderable current of very cellular lava seems to derive from this point, and clothes the southern slope of the mountain. The many fragments of granite imbedded in its scoriæ demonstrate that this eruption, though undoubtedly of a much later date than the basalt which forms the surface of the plateau, originated far beneath the calcareous freshwater strata which compose the hill. These strata contain veins of gypsum and pyrites, interspersed with bitumen or sulphate of barytes. Though generally horizontal, on some points they show considerable disturbance, and pass into peperino. The Puy de Coran is

---

* Bulletin XIV., p. 657.

worthy of much attention, as including the products of eruptions at distinct and very distant epochs; its scoriæ abound in octohedral crystals of oxydulous iron, in fiorite, and crystals of hornblende similar to those of the Puy Channat, but far larger. Some are nearly the size of the fist. They are occasionally perfect; but in general their exterior is rounded, and the angles blunted as if by a partial fusion.

87. *Puy de Cornon.*—Two remnants of a basaltic current are found upon this extensive and nearly flat calcareous hill; the remaining surface of which is mostly covered by a thick bed of boulders of granitic and volcanic rocks, and was obviously at one time the channel of the Allier, although elevated between 400 and 500 feet above this river's actual level. At its base the marly limestone passes on some points into peperino.

To the east of the Allier, between Pont de Château and Issoire, a considerable number of other plateaux and peaks of basalt may be observed, capping eminences both of the freshwater limestone and of the low granitic range between the Dore and the Allier, against which this formation abuts. These are the remaining segments of very ancient currents; their scoriæ have generally disappeared; and the points they derive from, as well as their connection, if any ever existed between them, are with difficulty to be traced. The space through which they show themselves is not very extensive, and may be considered as a short band enclosed between the Allier and the meridian of Mozun. Nearly all the remarkable hills of calcareous peperino described above as part of the freshwater formation of the Limagne are also found near these limits. The direction of this volcanic line seems to preserve a parallelism to that of the Monts Dôme.

The most remarkable of these basaltic remnants are the Puys Bénôit, Dallet (already noticed, p. 15), La Roche Noire, St.

Romain, Turluron, and St. Hippolyte, which are based upon the tertiary strata; and the Pics de Buron and Mozun, which rest in part on the primitive soil. They present no peculiar features, and therefore to describe them in detail would only be to repeat what has been said already of the similar rocks in the immediate vicinity of the Monts Dôme.

CHAPTER VI.

REGION II.—THE MONT DORE.

§ 1. General Outline of the Mont Dore.*

As yet we have found the primary soil concealed but by occasional masses of volcanic rocks, between which it crops out at no very distant intervals. But I have now to describe one of those mountainous excrescences which have covered its surface to an extent of many miles in diameter, and elevated themselves to a proportionate height above its level.

The Mont Dore, though not the most considerable of the three in bulk or extent, attains the greatest absolute elevation. Its highest point, the Pic de Sancy, is given by Ramond as 6258 feet, exceeding that of the Cantal by 128 feet. Its figure will be best understood by supposing seven or eight rocky summits grouped together within a circuit of about a mile in diameter; from whence, as from the apex of a flattened and somewhat irregular cone, all the sides slope more or less rapidly, until their inclination is gradually lost in the high plain around. Imagine this mass deeply and widely eaten into on opposite sides by two principal valleys, (those of the Dordogne and of Chambon,) and further furrowed by about a dozen minor water-channels, all having their sources near the central eminences, and directing

---

* The Mont Dore, anciently Mons Duranius, derives its appellation from the stream called *Le Dore* which rises on its summit, and is therefore improperly written *Mont d'Or*. See Ramond, *Nivellement des Plaines*: Mém. de l'Inst. 1815.

themselves from thence to every point of the horizon. You will then have a rude but not inaccurate idea of the Mont Dore.

It is barely possible that some mountain, not volcanic, may, by long isolation, or accidental circumstances, have assumed somewhat of this form, but the additional peculiarity which the Mont Dore and Cantal share with Ætna, the Peak of Teneriffe, Palma, and all other insulated volcanic mountains, is, that the rocks of which each is composed exhibit themselves in beds every way dipping off from the central axis, and lying parallel to the external sloping flanks. This singular disposition would induce us à *priori* to conclude these mountains to be the remains of vast volcanos. The idea is of course confirmed, when we discover on examination that they consist of prodigious layers of scoriæ, pumice-stones, and their fine detritus, interstratified with beds of trachyte and basalt, which bear the stamp of an igneous origin, and descend often in uninterrupted currents, till they reach and spread themselves over the platform around the base of the mountain.

It is true that no regular crater remains on this summit. It would be irrational to expect one in a volcanic mountain which exhibits so many other proofs of having been long and violently attacked by the agents of dilapidation since the extinction of its fires.

The fragmentary ejections of its vent have gone in great part to form the immense conglomerates that clothe its sides and accumulate at its foot. Its more durable productions, its lava-currents and some consolidated breccias, have more successfully resisted the wear and tear of ages, and their highest extremities still bristle in elevated peaks over a circus-like gorge, which occupies the very heart of the mountain, and was probably the site of its central crater, but which now, branching out into deep and short recesses, forms the upper basin of the principal valley, and the recipient in which two mountain rills, the Dore and Dogne, unite, at the source of the noble river which from thenceforward bears their joint names.

If the materials of a volcanic mountain were arranged in any sort of uniformity, the valleys which have reduced the Mont Dore to a mere skeleton would exhibit its constitution in the most satisfactory manner; but as might be expected, the sections they offer disclose only vast and irregular layers of tuff and breccias, mingled with repeated or alternating currents of trachyte, clinkstone, and basalt, and traversed by numerous dykes of the same rocks.*

The opposite sides of each excavation generally offer corresponding sections, the same beds being visible at similar heights on both declivities, but varying occasionally in thickness. This is universally the case in all the narrower gorges near the base of the mountain, where the diminished slope caused the lava-currents to increase in width as much as in length; and in these situations the same bed or series of beds often extends over a surface of many square miles, forming a succession of vast platforms, with a slight, and, towards their termination, scarcely perceptible declination.

On examining the currents which compose these distant

---

* Were the causes which occasion the activity of Ætna to cease, this volcanic mountain would before the lapse of many centuries assume the chief characteristic features of the Mont Dore. Even now, its sides are furrowed by deep and vast valleys produced by earthquakes and the rapid descent of torrents of rain. The beds of lava of different epochs may be seen forming numerous pseudo-strata one above the other, and corresponding on the opposite sides of these valleys; the most remarkable of which in this respect is that of Trisoglietto. See Ferrara, *Descrizione dell' Etna*, 1818.

It appears that the flanks of the Peak of Teneriffe are yet more deeply intersected by rents and ravines; and M. Escobar is said to have counted above 100 strata of different lavas and beds of pumice on the sides of the valley of las Guanchas, N.W. of the Peak. See too the description of the Islands of Palma and Madeira in Sir C. Lyell's Manual, ed. 1855, p. 498 *et seq.*

plateaux, they are found to consist of basalt, which has flowed on all sides to the distance of 15 and 20, and in some instances, on the east and north, of 25 or 30 miles from the central heights.* Though the continuity of some of these sheets of basaltic lava has been destroyed, we may remount many of them without meeting any interruption, till at no great distance from the summit of the group we arrive at a spot, which, from the torrefied and vesicular nature of the basalt, and the number of scoriæ and bombs still adhering to its surface, appears to be the source of the current, the vent from which it was expelled.

The plateaux of trachyte, on the contrary, rarely reach to such an extent, and few portions of them deriving from the Mont Dore are to be found without the limits of a circle of 10 miles radius. But what these currents lose in length they make up in height and width. The lavas of this class appear to have possessed an inferior degree of fluidity to those of basalt; probably owing to their inferior specific gravity † and greater coarseness of grain; and in consequence they have accumulated upon one another in prodigious volumes in the vicinity of the source. They thus become the most conspicuous if not the most considerable portions of the edifice which they have reared in common with the others. Trachyte constitutes nearly all the principal heights and central platforms of the mountain, while basalt rarely shows itself but on its outer slopes or in the lateral escarpments and at the bottoms of its valleys.

A few isolated fragments of basaltic currents may, however, be

---

* These dimensions are far from being unparalleled by the lavas of modern volcanos. Sir W. Hamilton reckoned the current which reached Catania in 1669 to be 14 miles long, and in some parts 6 wide. Recupero measured the length of another, upon the northern side of Ætna, and found it 40 miles. Spallanzani mentions currents of 15, 20, and 30 miles (*Voy. en Sicile*, i. 219); and Pennant describes one which issued from a volcano of Iceland in 1783, and covered a surface of 94 miles by 50! (*North Globe*, vol. i.)

† See Considerations on Volcanos, pp. 86, 92, *et seq.*

seen at a considerable elevation, and resting, without interposition of other substances, on some of the trachytic plateaux; while instances of so clear a superposition of trachyte to basalt are less common. Hence it has been supposed that the eruptions which produced the lavas of this latter sort took place after the cessation and final extinction of the volcano which gave birth to those of trachyte. But facts are far from warranting this supposition; and I shall have to adduce several examples of the evident alternation of the two rocks.

Nor, indeed, are they always to be accurately distinguished from each other. Amongst the infinitely diversified varieties of trachyte to be found at the Mont Dore, where no two beds are alike, and even the same frequently changes its aspect in a considerable degree, several have the laminar structure, foliated texture, and scaly grain which characterizes clinkstone. When these, as they occasionally do, contain a large proportion of augite, they approach closely to, and in fact are undistinguishable from, basalt. Where the texture is not scaly, but the quantity of augite considerable, the trachytes often assume the exact appearance of some of the recent lavas of the Monts Dôme, are extremely cellular, of a dark grey colour, and crystalline texture. The trachyte used as a building-stone at Mont Dore les Bains is almost identical with the lava of the Puy de Nugère quarried at Volvic.

The whole quantity of *Fragmentary* matters ejected by the principal and subsidiary vents of the Mont Dore must once have fully equalled that of its lava currents; but the loose nature of these conglomerates has exposed them, of course, to more speedy destruction. The volume of those which remain is, however, prodigious. They in turns rest upon, support, and envelop the massive lava-rocks of every kind. They are found at every distance from the centre of eruption; sometimes spreading into wide plateaux, at others filling the bosom of mountain hollows, like the masses of drift snow left in a hilly country by a brief thaw.

These conglomerates are susceptible of a division into two species, according as either class of volcanic products predominates in their composition.

Some consist wholly of triturated pumice, in which the fine silky filaments of this substance are to be recognised, as well as a few crystals of felspar. This occurs either loose and arenaceous, or consolidated by an intimate mixture with water into a yellowish-white tuff of a certain consistence, resembling the tufa of the Phlegræan fields, near Naples: occasionally it has a lamellar structure, and has been sold in commerce for tripoli. In general, however, this pulverulent substance envelops various-sized fragments of trachyte, basalt, and granite, forming a tufaceous conglomerate. As these coarser materials predominate, a complete breccia is the result, in which the fragments are immediately in contact, or separated by occasional interstices, or finally agglutinated by a cement, either of tuff or of iron-rust, apparently derived from the partial decomposition of the fragments themselves, which are in these instances for the most part of a highly ferruginous basalt, and in this condition it resembles the peperino of the Campagna round Rome.

M. Ramond, in describing these conglomerates,* very justly remarks that their disposition excludes the idea that the waters either of the sea or of inland lakes have had any share in their arrangement;† and he imagines them in consequence to have

---

* Mémoires de l'Institut, 1815.
† The greater number of the extinct volcanos of Italy having been apparently submarine, it follows of course that the loose matters ejected by them should have been deposited, spread out,

fallen from the air upon the spot they still occupy after their projection by the explosive force of the volcano.

But it will be difficult for those to coincide fully in such an opinion who have remarked that the great proportion of these deposits are found, not in the neighbourhood of the crater, but in massive and partial accumulations at the foot of the mountain, extending frequently in a direct line to a very great distance from its centre, without altering their character or suffering any corresponding diminution in the size of their fragments.

The conglomerates which compose the plateaux of Pardines, Nechers, and Polagnat, for example, as well as those capping the Puy de Monton, are too unequally distributed, too distant from the focus of eruption, to have been formed in the manner supposed by M. Ramond. At the latter point, which is exactly 20 miles in a straight line from the central summits of the Mont Dore, where certainly the vent which ejected them was situated, I have observed massive blocks of very compact trachyte and trachytic obsidian, measuring more than a cubic yard in bulk. It is incredible that fragments of this magnitude should have been urged to such a distance merely by the projecting power of the volcano, a power which is always exerted in a direction very nearly if not quite perpendicular. Nor could the prevailing winds, which, M. Ramond remarks, set that way, have had any notable effect upon bodies of so vast a weight. We must, there-

---

and levelled at the bottom or on the shores of the sea: and hence the even and stratified tuffs of the western coast of that peninsula, from Naples to Civita Vecchia. See Brocchi, *Suolo di Roma*, pp. 180, &c. &c.; and Breislak, *Voy. en Campanie*.

But the case is very different with the volcanos under consideration, which raged upon the summit of an extensive table-land, elevated between 3000 and 5000 feet above the actual level of the sea; and whose earliest eruptions appear to have commenced not only long after this land had emerged from the ancient ocean, but at a period when the fresh-water lakes of the district had almost, if not entirely, terminated their deposits.

fore, of necessity admit the agency of some other cause in these instances; and various facts tend to prove that the rapid descent of water from the summit of the mountain at the period of its eruptions co-operated with the fall of ejected stones and ashes in the formation of these conglomerates.

These facts are principally the following:—

1. The deposits of this nature, which advance to a distance from the mountain, occupy a few large valleys, obviously excavated in the granite or the freshwater strata previous to their deposition, and are not found in the wide intervals separating these valleys, as would certainly be the case had they resulted from the uniform dispersion of loose fragments ejected by the volcano.

2. Their participation in the character of alluvia is evinced in the consolidation by water and partial stratification of some of the tuffs; in the occasional layers and heaps of sand and gravel accompanying them, the trunks and logs of trees and the remains of land animals frequently found imbedded in them.

3. The greater number of fragments composing the distant conglomerates have their angles broken or rounded, and the largest are completely bouldered.

4. Immense fasciculi of basaltic prisms are occasionally seen enveloped by them. The prisms are not separated from each other, nor their angles injured. These rocks cannot have been launched thus from the crater, but must have been detached from neighbouring currents by some violent and erosive attack.

5. The conglomerates found within the limits of the freshwater formation contain rounded fragments of limestone. These could not have been thrown up by the volcano, which broke out on the *granite;* but it is easily seen that the calcareous strata may have been torn up by torrents hurrying along and depositing on their route the beds enclosing these blocks.

They must not be confounded, however, with the ordinary drift deposits of any mountainous country resulting from the abrasion of its rocks and excavation of its valleys,—from which they are distinguished not only by their volume, but by the currents of trachyte, clinkstone, and basalt which cover, support, and penetrate them in all directions, and were, therefore, certainly of contemporary formation.

The descent of aqueous deluges down the sides of great habitual volcanos during the occurrence of their eruptions, is well known to be a common phenomenon,[*] and to be owing to one or more of three distinct causes :—viz.

1. The sudden melting of snow on the summit of the mountain, either by falling showers of red-hot scoriæ, or the contact of a disgorged current of lava.

2. Prodigious storms of rain which usually accompany or succeed to volcanic eruptions, and have been satisfactorily attributed to the condensation of the immense volumes of aqueous vapour evolved from these vents during their activity, and which constitute in fact the main agent in their phenomena.

3. The vast bodies of water, probably the contents of internal crater-lakes, which *trachytic* volcanos, particularly those of America,[†] are known occasionally to eject, and which, when mingled with the ashes and lapilli caught up in their progress, or brought with them from the interior of the volcano, have been sometimes called " currents of mud," but must not be confounded with the product of those pseudo-volcanic explosions of hydrogen gas seen at Macalouba, Modena, in the Crimea, &c. &c. It is indifferent which of these various phenomena we suppose to have

---

[*] See Considerations on Volcanos, page 158.

[†] See Humboldt, *Tableau Physique*. Daubuisson, *Geogn.*, vol. i. p. 181.

Breislak, *Inst. Geol.*, vol. iii. § 641. Scrope on Volcanos, p. 160. Lyell, Principles, ed. 1853, p. 412.

operated in the present case: possibly all may have been in turn or at once brought into action.*

If we consider the effects that would naturally be produced by the sudden rush of large bodies of water down the sides of an elevated volcano like the Mont Dore, at its moments of eruption, sweeping away all the loose materials surrounding its crater, as well as all they meet in their descent, tearing up the flanks of the mountain, and overwhelming the plains or valleys around, we shall find them fully equal to the formation of the conglomerates that now occupy us, and which, whether filling hollows in the mountain's sides, or deep and extensive valleys to a considerable distance from its base, are found to exhibit just that chaotic confusion which we should expect from such a mode of production, and which no other can sufficiently account for.

The Puy de Monton is perhaps the point furthest removed from the Mont Dore at which these eluvial products are found in abundance. I have mentioned its direct distance from the Pic de Sancy as twenty miles; the difference of level is about 4400 feet; a difference which affords an average fall of one foot in twenty-four, certainly sufficient to give to any great body of water the impetus required for transporting blocks of the size already noticed as occurring there.

---

* The Mont Dore is covered every winter with great quantities of snow, which seldom disappears till late in the summer. There were some considerable patches remaining in the hollows near its summit when I visited it in the beginning of *September*, 1821, as well as at the time of my last visit in August, 1857. With respect to the occurrence last instanced, as taking place among the American volcanos, it is worth noticing that, like those of France under consideration, they are principally *trachytic*. The fine ashes of trachytic volcanos unite with water into a tenacious clay, which circumstance often occasions the creation of lakes within their craters during intervals of tranquillity. These lakes either burst their banks by their increasing weight, or are let loose by the next eruption, in either case rushing in a tremendous eluvial debacle on the lower grounds around the mountain. The trass of the Rhine volcanos was no doubt produced in this manner.—See Considerations on Volcanos, p. 159.

I have been anxious to substantiate the mode of formation of these conglomerates, which has not been always satisfactorily explained, on account of their forming so prominent a feature in the trachytic formations of central France; constituting perhaps, in the three great volcanic groups of this class, a full moiety of their whole mass, and appertaining exclusively to these, since no similar rock is to be found in connection with the basaltic products of the more recent or minor volcanic vents.

§ 2. Structure of the Mont Dore.

To describe the constitution of the Mont Dore in detail, rock by rock, as far as it can be observed, would swell these pages to too great a bulk; I therefore content myself with sketching its principal masses, pointing out whatever is most remarkable in them.

Let us, for this purpose, imagine ourselves for a moment on the Pic de Sancy, a pyramidal rock of porphyritic trachyte, and the highest point of the whole mountain.* Connected with this by intervening ridges, rise on each side similar craggy knolls of the same substance, more or less rounded by weathering, and partly covered with vegetation. One of them, the Puy Ferrand, almost equals the Pic de Sancy in elevation. These two most prominent heights overlook, on our right and left, two deep amphitheatral basins, one opening to the north, encircled with a range of perpendicular precipices in which the different sources of the Dordogne unite; the other to the north-west forming the gorge of Chandefour at the head of the valley of Chambon. On the side opposite to these hollows each eminence gives rise to an

---

* The localities mentioned in this sketch of the Mont Dore may be observed in the Departmental Map of the Puy de Dôme of 1845. But see the bird's-eye view of the valley of the Dordogne in Plate VII.

inclined plane with a gradually decreasing slope, perhaps broken at first into three or four step-like projections, one above the other, and by degrees widening, as it descends, into vast platforms,* which with few interruptions reach the base of the mountain, and prolong themselves to some distance over the adjoining country.

The rock which composes these platforms is, almost without exception, trachyte, and the general divergence of all the principal trachytic currents from one spot leads us to presume them the produce of a single habitual vent, occupying this central situation. The structure of the heights surrounding the gorges mentioned above confirms this supposition. They consist of various and often-repeated beds of trachyte, exhibiting in their confused arrangement and scoriform parts features which characterize the proximity of a centre of eruption. On descending we observe rocky masses of conglomerate alternating with, or leaning against them. Of this nature is the high ledge whence spring the two rivulets Le Dore and La Dogne, and which unites the four chief central eminences, the Pic de Sancy, the Puy Ferrand, the Pan de la Grange, and Cacadogne. Immediately beneath the Cascade of the Dore, this rock contains alum and sulphur in such abundance as to repay the cost of considerable works for their extraction. To the west, two deep gorges, called

---

* The high lands of the Mont Dore, as well as of the Cantal and Mezen, are too much exposed and too elevated for cultivation, but are clothed with an unlimited succession of rich and widely-spreading pasturages. The constitution of the soil is sometimes thus concealed for a considerable space. It is, however, always disclosed at intervals in the steep banks of the water-channels which drain the plateaux. With the exception of a few fir forests in the highest regions, little wood is to be found out of the valleys; and this peculiar disposition of surface and vegetation gives to the country an entirely different aspect according as it is viewed in its valleys or on the intervening flat-topped heights. The former are beautiful, luxuriant, and populous; the latter naked, dreary, and almost uninhabited.

Les Vallées de l'Enfer and de la Cour, are completely scooped out of the same rock. It has a torrefied aspect, and consists of fragments of trachyte and basalt, both compact and cellular, united into a breccia sometimes by a ferruginous, at others by a pumiceous cement. In the former case the compound is exceedingly solid and durable. It is penetrated by several vertical dykes of porphyritic trachyte, of a dark colour, vesicular in parts, and divided into regular columns, generally at right angles to the walls of the vein. A narrow and low partition, called Les Fernes, which separates the gorges of la Cour and l'Enfer, principally consists of such a dyke. The enclosing breccia has disappeared on one side, but remains on the opposite. In the former ravine are two or three others entirely denuded, and exactly imitating the ruins of Cyclopean walls; the extremities of the prisms, which are laid horizontally, showing their polygonal surfaces on each side. The steep face of the Puy de l'Aiguiller, terminating the Vallée de l'Enfer, exhibits three or four similar dykes traversing vertically its whole height (900 feet), and by their *needle-like* peaks, which rise almost to the height of the Pic de Sancy, occasioning its name.

Immediately opposite to the Vallée de la Cour, on the east side of the valley of the Dordogne, is a deep ravine separating two craggy cliffs called Cacadogne and Le Roc de Cuzau. It is strewed with colossal ruins from the rocks above, which consist of conglomerate enveloping various currents of trachyte and basalt mingled in strange confusion. Among the blocks scattered below, and belonging to these currents, are many of a trachyte approaching to obsidian, of resinous lustre and fracture, and a black colour, enclosing numerous large crystals of felspar (pitch-stone porphyry); and another rarer variety, compact, hard, of a brick-red colour, with something of the resinous gloss of pitch-stone, enclosing opaque crystals of felspar exceedingly hard and of a waxy look. The *roc Barbu*, an insulated and shapeless rock springing from the middle of the ravine, presents a cluster of diverging columns of an extremely dense, heavy, and hard basalt, enclosing large and regular crystals of augite and olivine.

Such is the nature of the area overlooked by the central trachytic summits; and in these features it is easy to recognise the traces of a vast and ruinous crater, not very dissimilar to the picture exhibited at the present moment* by the recent crater of Vesuvius, torn through the bowels of the mountain by the eruption of 1822; whose abrupt and precipitous escarpments, like those of the gorges just described, are composed of a conglomerate of scoriæ and volcanic fragments, enveloping horizontal beds of lava, and penetrated by numerous dykes of the same substance, mostly vertical, and separating into horizontal prisms.

We seem then authorized to conclude the principal vent from which issued the great formations of trachyte of the Mont Dore to have been situated in the immediate vicinity of the upper basin of the valley of the Dordogne.

There is, however, no reason to believe that this vent was productive of trachyte alone; the numerous fragments and dykes of basalt enclosed in the surrounding conglomerates, together

---

* This was written in 1823. It must be remembered that the central crater of every habitually eruptive volcano is liable to be alternately emptied by paroxysmal explosions, and refilled subsequently by minor eruptions from within its area. (See a paper on the Formation of Craters, &c., 'Geolog. Proc.,' April, 1856.) The latest eruptions of an extinct volcano may have belonged to one or the other of these classes, and consequently may either have left a vast chasm (or crater) in the heart of the mountain, or in its place a solid dome or cone composed of the products of the preceding eruptions in the shape of fragmentary accumulations penetrated by dykes and beds of lava in a state of chaotic confusion. Or the last explosions may have only blown up portions of such a cone, occasioning one or more of those circus-like hollows so often found towards the centre of such mountains, and of which we have examples both here and in the Cantal.

with the direction and slope of the neighbouring basaltic beds, tend rather to prove the contrary; and it has been already noticed that many of the currents of this rock, so conspicuous on the skirts of the mountain, may be traced to points of eruption near the central trachytic plateaux, and therefore at no great distance from the principal crater.

If, quitting the circus where the waters of the Dore and Dogne meet, we follow the torrent which bears from this point their united names,* we find perpendicular cliffs, composed of repeated beds of trachyte, bounding the valley on each hand. To the left a deep ravine has laid bare the face of a steep rock called Le Cliergue, and disclosed the out-cropping of five or six immense currents of trachyte, separated from one another by layers of tuff and decomposed pumice. The lowest of these enormous beds are quarried for the building-stone of the neighbouring baths; and the rock of which they consist is excellently fitted for the purpose, separating readily into rude prismatic masses, being extremely cellular, working well under the tool, and resembling strongly the lava of Nugère or "Pierre de Volvic." Its colour is bluish-grey; it has few apparent crystals of felspar, and is highly sonorous. The upper beds are of another variety, of a lighter colour, and contain larger and more numerous crystals of felspar.

A similar order of beds is seen on the opposite side of the valley. Their upper surface slopes rapidly from the summit of the Roc de Cuzau, forming a plateau called La Durbise, which is prolonged below in the Plateau de l'Angle, immediately above the village of the Baths. Of this the opposite Plateau de Rigolet obviously once formed part, though now separated from it by the excavation of the valley of the Dor-

---

* See Plate VII.

dogne. The height, slope, and direction of each correspond, as well as the beds of which they are composed. The upper surface of both is occupied by a thick current of trachyte, the section of which on each side of the valley presents a long range of irregular columns.

8. Cascade above Mont Dore les Bains.

A rent worn through this and some of the inferior beds by a waterfall called *La Cascade du Mont Dore*, at a short distance above the Baths, exhibits them in the following succession:—

1. (Beginning from above.) A bed of porphyritic trachyte, 160 feet in thickness, forming the floor of the plateau to its extremity. The colour of this stone varies from greyish to bluish-white. It bears a considerable resemblance to the rock of the Puy de Dôme, like which its fissures are sometimes lined

with large blades of specular iron. It is highly porous and contains numerous large crystals of glassy felspar, mica, and acicular hornblende enveloped in a very loose and coarse-grained base, evidently composed of the detritus of these crystals, together with grains of oxydulous iron and perhaps also augite. It resembles strongly the trachyte of *Drachenfels.* This rock often encloses spheroidal masses, of a darker colour and more compact than the surrounding substance, penetrated by interlaced acicular crystals of hornblende, which recall the similar nodules in the trachyte of the Euganean Hills and those which occur in some granites and porphyries.

2. This trachyte is superposed to a thick bed of arenaceous tuff evidently belonging to the stratum above. In its upper part are many loose crystals of glassy felspar, large and perfect, generally double. Their outside is mealy; their interior carious, often presenting longitudinal fibres separated by equal interstices. This appearance seems to have been produced by a partial fusion, in the manner of the filaments of pumice. Similar crystals are met with in the highly porphyritic trachytes of Ischia.

3. Columnar clinkstone, passing into basalt,* highly schistose in parts, of a dark slate-colour, imbedding numerous small crystals of augite and glassy felspar. It is slightly translucent at the edges.

---

* It has been proposed that the presence of olivine should be established as characteristic of basalt, which in all its other characters is often liable to be confounded with clinkstone and trachyte. But this is impossible, since many typical basalts are to appearance devoid of that mineral. All that can be said is that the very feldspathose lavas are usually recognised as trachyte, the most augitic and ferruginous as basalt; the trachytic rocks, when compact and schistose in structure, being called clinkstone. In an article published in the Journal of the Royal Institution for June, 1826, and containing a proposed conventional arrangement for the description and nomenclature of volcanic rocks, I ventured to propose an intermediate genus to include those rocks which partake both of the characters of trachyte and basalt, to be called Greystone (Graustein), from their prevailing, indeed invariable, tint, and I still think some such denomination desirable.

4. Breccia of scoriæ and volcanic fragments with a cement of tuff.

5. Thick beds of amorphous basalt, varying its characters. In some parts of a dark slaty grey colour, compact, sonorous, and containing small crystals of felspar: in others, reddish-brown, heavy, and close-grained, but studded with large elliptical cells, the interior of which is generally lined with small mammillæ of hematite, and the rock is throughout exceedingly iron-shot.

This bed of basalt and its accompanying breccias make their appearance also below the hood-shaped rock of trachyte called *Le Capucin*, on the opposite side of the valley, and in the valley of La Scie, supporting that side of the Plateau de Rigolet. The columnar basalt of the *Cascade de la Querilh*, at the extremity of the Plateau de l'Angle, which appears to crop out from beneath the superficial beds of trachyte and tuff, probably belongs to the same lava-current.

6. White pumiceous tuff enveloping a few fragments of granite, basalt, and trachyte. This bed is traversed by two or three nearly vertical veins which have evidently proceeded from the basalt above; their lower extremities are seen to terminate in the tuff. This and their insignificant proportions oppose all idea of protrusion from below.*

---

* This remarkable natural section presents itself so obviously to all the visitors of the Mont Dore, lying at the distance of scarcely a quarter of a mile from the village of the Baths, and under the Cascade which is the chief *lion* of the place, that I own myself wholly unable to comprehend how M. Beudant and other French geologists could in the face of it have denied the superposition of trachyte to tuff or basalt. At the Ravin des Egravats, higher up the valley, where a vast landslip took place a few years since, and disclosed the structure of the mountain in a steep cliff half a mile in length, a bed of decided basalt (perhaps the same) also underlies the mass of conglomerates which supports the great upper platform of trachyte. On the whole it may be said that basalt is as frequently found to underlie trachyte as the reverse. It is, I think, impossible to admit the general priority of the latter

Nearly parallel to the valley of the Dordogne on the west, and divided from it by the Plateau *de Rigolet*, is that of a torrent called *La Scie*. The mountain that rises beyond is capped by an enormous current of porphyritic trachyte, which forms the floor of an elevated plateau called *Bozat*, the prolongation of that of Le Cliergue, and which evidently was propelled in a massive stream from the vicinity of the *Puy de la Grange*, an eminence impending over the *Vallée de la Cour*. Still further to the west, the valley of Vandeix has divided this plateau from that of the forest of *Charlanne;* but the perfect correspondence in slope and constitution of this and a suite of other high plains succeeding to it in the same direction, the last of which is crossed by the road from Clermont to Aurillac, proves them all to have been originally united in one of the most extensive and voluminous trachytic lava-currents of the Mont Dore. The rock of which it is composed is porous and in parts highly cellular; the imbedded crystals of felspar numerous and large, attaining even *four* inches in length. A flat-topped range called *Chamablanc* branches off from that of Bozat towards the north, but at a much lower level, and a bed of basalt observable on its surface and sides bears every appearance of cropping-out from below the

---

to the former class of lavas which the French geologists usually assume as a fact. Another equally to me unaccountable opinion of M. Beudant is, that none of the trachytes of Mont Dore have the form of lava-currents or sheets; which is, on the contrary, their prevailing disposition, though their bulk or thickness is so considerable as to deceive a hasty observer. He admits, it is true, that there are *highly feldspathose lavas in the Mont Dore analogous to trachyte*, which assume the form of currents (*coulées*); but adds that the *true trachytes* are never found disposed in this manner.—The real fact is, that it is utterly impossible to distinguish, *mineralogically*, either in hand specimens, or on a larger scale, feldspathose rocks which are disposed in thick sheets and currents from those which have a somewhat more massive and bulky figure. M. Beudant, therefore, could only intend to confine the term Trachyte to a rock of a particular figure and of peculiar geological relations,—a construction he is not warranted in putting on a term generally received in a mineralogical sense.—See *Scrope on Volcanos*, p. 94 *et seq.*

trachyte and tuffs of the upper plateau, though the débris which conceal the junction of the two rocks render it difficult to ascertain the fact.

On the opposite, *i. e.* the eastern side of the valley of the Dordogne, and above the Plateau de l'Angle, is seen another embranchment of porphyritic trachyte of still larger proportions than the one just described. It appears originally to have constituted a massive and elevated ridge (similar to that of which we have seen so considerable a segment preserved in the Plateau de Bozat), deriving from the central summits of Cacadogne and Cuzau, and directed from south to north with a progressive inclination. Subsequent denudation, acting with great effect on a rock which yields so easily both to decomposition and abrasion, has reduced it to an irregular chain of round-topped hills, closely united by their bases, and gradually decreasing in height as they recede from the central summits. The principal of these go by the names of the Puys de l'Angle, Hautechaux, Barbier, La Tache, Poulet, Baladou, l'Aiguiller, and Pessade. At the last of them the trachyte abruptly terminates, exactly in face of the Monts Dôme; the interval is covered by basaltic currents, most of which appear to me to proceed from *beneath* the trachyte, and to have flowed from the central vent of Mont Dore.

The plateaux also which descend towards the east from the foot of this group, as well as those on the southern side of the valley of Chambon, which take their rise immediately from the Puy Ferrand, are chiefly of basalt. They spread with a gradual slope, in wide and uniform sheets, over an immense tract of country; at first but slightly furrowed by the mountain torrents, further on completely penetrated to the granite beneath; and, as we descend still lower and the valleys deepen, cut up into long strips which line the margins of every gorge with massive columnar ranges, and project in the form of flat-topped promontories into the plain of the Limagne. These terminate at the Allier, but on its further bank rise some few isolated cones of the same basalt, which mark the original extent of the currents, and prove that the river had not excavated its actual bed at the epoch of their descent. They are almost everywhere accompanied by beds of conglomerate equally derived from the central volcano, and probably for the most part drifted down to their present position by aqueous debacles, such as were referred to in a former page.

The two principal valleys that now drain this vast inclined plane, viz. those of Chambon and Besse, partly excavated through granite, and partly through the freshwater formation, but everywhere bordered by impending ranges of basalt and its conglomerates, are exceedingly interesting; and not the less so from the circumstance that the bottom of each has been occupied by a current of lava belonging to eruptions of a very recent date.* The upper basin of the first-mentioned valley exhibits in the overhanging cliffs porphyritic trachyte with conglomerates, resting on granite † and supporting basalt; the trachyte terminates above the village of Chambon, but the tuffs and breccias accompany the basaltic currents all the way to

---

* See Plate IX.

† This is the highest point at which the granitic substratum of the Mont Dore shows itself, viz. 3714 feet from the sea. It re-appears on the north-east of the mountain below *Murat le Quayre*, in the valley of the Dordogne, at the absolute elevation of 3271 feet; and on the south east near Chastriex, of 3422 feet. If we take the mean of these three (3469 feet) for its probable elevation beneath the central summits of the mountain, we shall have 2748 feet as the depth of the volcanic products alone on that point, and above 1400 feet for their average thickness throughout a central circle of $3\frac{1}{2}$ miles radius.

This volume, however considerable, is far inferior to that of the Cantal, perhaps scarcely exceeds that of Vesuvius, and sinks into nothing when compared with the colossal bulk of Ætna or the Peak of Teneriffe, which consist principally of volcanic matter from the level of the sea, and indeed from a great depth below this, to the height of 11,000 and 12,000 feet: or the still more stupendous trachytic formations of the Andes and Cordilleras.

the Allier, showing themselves at intervals throughout both valleys in prodigious accumulations, as in the Dent du Marais near the Lake Chambon, and the plateaux of Pardines and Nechers.

They universally contain blocks of every variety of trachyte and basalt, fragments of granite, pumice-stones, scoriæ, &c., and a large proportion of titaniferous iron in their sandy detritus. Masses of limestone occur in them where they cover or rest against the freshwater strata; and in similar circumstances the basalt has sometimes its cellular cavities filled with calcareous infiltrations. At the Montagne de Laveille, near Chidrac, some highly amygdaloidal portions occur, in which arragonite and carbonate of lime form nearly two-thirds of the mass. It is immediately beneath, or occasionally intercalated among, these tuffs and breccias that the celebrated bone-beds of Mont Perrier occur, in which Messrs. Croizet, Bravard, and Pomel have detected numerous mammalian remains belonging to several distinct assemblages of species, which the two former naturalists refer to successive tertiary epochs.[*]

The Dordogne, which for the first three or four miles of its course flows nearly from south to north, makes a sudden bend to the west a short distance below the village of the Baths, leaving to the right a massive portion of high table-land which exactly fronts the whole of its upper valley, and was perhaps originally separated from the central heights by some violent explosions, while the subsequent excavation of the channel of the Dordogne has widened the breach.

The base of this mountain consists of various conglomerates,

---

[*] See Sir C. Lyell's Manual, p. 552, ed. 1855; and Quarterly Journal Geol. Soc., vol. ii. p. 77. Also the Catalogue of Organic Remains of Central France in Appendix *infrà*.

enveloping beds of basalt; above these a band of clinkstone may be traced across the whole of its eastern and northern faces, surmounted by porphyritic trachyte—if indeed these two rocks do not, as might from many circumstances be suspected, pass into each other. Finally, the upper surface of the plateau exhibits more recent currents of basalt, which appear to have had their origin there.

Trachyte, however, occupies by far the most conspicuous place amongst these rocks, constituting the Puy Gros, an enormous flat-topped boss on the eastern summit, and descending from thence to the west in a wide unbroken platform, though occasionally covered by basalt, as far as the village of La Queuille, where the current terminates in a range of gigantic six-sided columns, some of which I observed to be not less than 15 feet in diameter; their height is not proportioned to so great a bulk, rarely exceeding 30 feet. The rock which composes them is a dark-coloured trachyte approaching to basalt (greystone), imbedding felspar and augite crystals, and exceedingly cellular. Its largest cavities often contain radiated crystallizations of arragonite.

Clinkstone, or the laminar and scaly species of trachyte, predominates to the north of the Puy Gros, where a thick bed of it seems to have been separated into detached segments by the torrent which flows from thence to Rochefort. The largest of the masses thus isolated are the Puy de Loueire, and the Roches Sanadoire and La Tuilière. At the first the phonolite is divided into compact tables; at the two last rocks, into very regular prisms. Those of Sanadoire are entangled into fantastic groups, in one spot diverging so regularly from a common centre as to resemble a circular fan. The prisms of La Tuilière are vertical or nearly so, and schistose, splitting into thin laminæ, which at the northern extremity of the rock are at right angles

Plate VIII.

29 AP 58

to the axes of the prisms, but acquire gradually an increase of inclination, till at the other end their obliquity is such that their planes make an angle of but 15 or 20 degrees with the axes, and the agency of gravity co-operates with that of the weather in separating them: the rock on this side is in consequence completely in ruins. These plates are used as roofing-slates throughout the neighbourhood, and hence the name of La Tuilière.

This clinkstone contains occasional crystals of felspar, and assumes in parts so much of the aspect of trachyte, that I am inclined to suppose it but an accidental variety of this rock, more particularly as the bed to which it belongs appears to merge in the great trachytic currents of the Puy Gros on one side, and the Puy de l'Aiguiller on the other.

The volcanic nature of the Roche Sanadoire was at one time strongly contested by naturalists, who examined the individual rock or its specimens alone, without consulting its evident connections with those around. But the frequent occurrence of cellular portions, and occasional scoriæ imbedded in its prisms, might have sufficed to convince even these sceptics of its igneous origin. Dr. Weiss of Leipsic discovered grains of haüyne in this rock, but they are almost microscopic, and very rare.

The basalt of the high plateau now under description appears to have been produced by repeated eruptions from a vent to the north-west of the Puy Gros, the site of which is marked out by two elevations entirely of scoriæ, called Chantouzet and Le Cros de Pèze.

Its currents have flowed both to the east and west, forming on one side the margin of the lake Guèry, on the other descending into the valley of the Dordogne, and exhibiting many prismatic ranges along its banks, as far as St. Sauve. They are accompanied by breccias, and may be observed at Murat le

Quayre, le Roc de la Montilhe, and other points, to be superposed to a thick bed of scoriæ as fresh in aspect as those of any recent volcano.

A still more extensive series of basaltic currents stretches from the foot of the Puy de l'Aiguiller towards the north, reaching to the distance of 16 miles along the banks of the Sioule, and covering great part of the granitic plateau of the Monts Dôme. Here, as elsewhere, they are found in association with the conglomerates, which accompany them in vast quantities almost to the end of their course. A wide basin parallel to the granite range of the Monts Dôme appears to have been once completely filled by these united volcanic products, of which immense portions still remain there. At Polognat the tuff consists principally of very white and silky pumice-stones evidently stratified by water. The conglomerates of this neighbourhood not unfrequently enclose trunks of trees, in which a complete passage may be traced from simple carbonization to the state of jet. They also contain grains of sulphur between their fibres.

The variety of basaltic currents which have taken this direction is remarkable. Near Rochefort they form repeated and enormous beds, divided into very regular prisms or tables. The latter modification of structure is also exhibited in perfection on the border of the valley of the Sioule opposite St. Bonnet, where tabular masses are often extracted ten feet long and proportionately wide, with a thickness of but three or four inches. Their surfaces are smooth and level; they are exceedingly elastic, and ring upon being struck exactly like a plate of cast iron. The basalt of which they consist approaches to clinkstone, is fine-grained and highly crystalline, perfectly compact, of a light slate-colour, with a tinge of green, and totally free from any imbedded crystals, though an imperfect agglomeration of augitic particles into globular concretions is betrayed by weathering. Near Villejacques, in the same valley, a current makes its appearance from amongst the surrounding tuff, the basalt of which encloses brilliant laminar crystals of hornblende sometimes two inches in length, and granular ones of olivine. It decomposes into minute angular spheroids. Another neighbouring current is formed of a very fresh and cellular basalt, resembling some lavas of the chain of puys; another is highly impregnated with iron, black, dense, and heavy.

The south-western face of the Mont Dore remains to be noticed. It presents a more uniform and smoother slope than the others. The currents of trachyte have proceeded but a short distance in this direction from the central heights. They constitute two or three salient masses composed of a porphyritic rock more or less porous, and in the vicinity of the supposed crater even scoriform and tinged of a deep red colour. Basalt on the contrary is extremely abundant on this side. It descends in extensive plateaux, from the extremities of the trachytic beds, and accompanied by a breccia of scoriæ on which it generally rests. Wherever these plateaux have been channelled by torrents, their sections offer ranges of columnar prisms of the greatest regularity; as at Chastreix, La Tour d'Auvergne, &c. At the latter spot the prisms are jointed, fitting together by means of alternately convex and concave bases; when broken they disclose a cylinder of compact and black basalt within a prismatic case of a lighter colour and looser texture.

The limits of the Mont Dore on the south are not very definite. The prolongation of its base meets that of the Cantal, and unites with it in forming a high and massive table-land, which divides the waters of the Dordogne and Allier, and is known by the name of Les Montagnes de Cézallier. A scanty vegetation covers the surface of this desolate and extensive district. Its soil chiefly consists of primary rock, until the limit of the department of the Cantal is reached, where basalt re-appears again in abundance.

Gneiss and mica-schist support several basaltic plateaux along the eastern slope of this district, and at a considerable elevation, as near Anza, Apcher, &c.: but it is probable that this high primitive ridge, by opposing an impediment to the progress of the lavas of the Mont Dore towards the east, was the cause of their comparative absence in this direction. On some points, as in the neighbourhood of Ardes, the basalt is seen to rest on the tertiary green and red clays, and sandstones.

To the west the inclination of this elevated plain towards the Dordogne is gradual, and its surface strewed with huge boulders of basalt and primitive rocks, attesting the ravages of the pluvial torrents from either mountain, and the frequent shifting of their beds.

§ 3. Recent Volcanic Eruptions of the Mont Dore.

It has been already mentioned that volcanic cones and currents, the products of eruptions comparatively recent, are found within the limits of the Mont Dore. They present exactly the same characters as the chain of puys of the Monts Dôme, belong apparently to the same æra, and have burst forth on the prolongation of the same line.

Beginning from the north, the first of these cones bears the name of the Puy de *Tartaret,* and is distant between three and four miles from that of *Montenard,* the last in this direction of the puys described above as belonging to the Monts Dôme.

It has been thrown up in the middle of the valley of Chambon; and, by obstructing the rivulet which flowed there, gave rise to the formation of an expanse of water, called the Lake of Chambon,* which appears from the flat and alluvial surface of the plain above to have been originally much more considerable, and to have retreated gradually in proportion to the excavation of the channel through which its excess is now discharged.

Although the hills on each side are loaded with the ancient volcanic currents of the Mont Dore, granite shows itself continually at their bases; and it is through this rock that the eruption of Tartaret has taken place, displacing and tearing up a superficial bed of basalt, which appears to have pre-existed on the spot.

The cone is composed throughout of loose scoriæ, lapilli, and fragments of granite. It has two deep and regular bowl-shaped craters, separated by a high ridge, and each broken down on one side. They have furnished together a very copious lava-current, which at first spreads over a wide and level surface, then, contracting itself as the valley becomes narrower, occupies the channel of the former river, and follows all its sinuosities as far as Nechers, below which it terminates at a distance of thirteen miles from its origin.

It is an interesting study to follow, through so long a course, this current of recent basalt from its source within an indisputable volcanic cone, along the bottom of a valley cased partly in granite, but whose sides are everywhere fringed by sections of more ancient currents, which have flowed in the same direction; to observe the analogy of these basaltic formations of different epochs, their perfect parallelism and almost equal extent, at least in one direction; their frequent similarity of structure; for in many parts, particularly near Champeix and Nechers, the more recent basalt is divided into regular columnar prisms;

---

* See Plate IX.

their common mineralogical characters, and the scoriform parts that accompany each.

There are few known countries in which such admirable juxtapositions of ancient and recent volcanic products are to be found; and hence, perhaps, the repugnance long exhibited by those who had only examined the extremes of the series at remote spots and distant periods, to admit their identity of origin. Here Nature herself has brought together the objects of comparison and placed them at once before the observer's eyes, as if for the express intention of teaching him their common mode of formation. In the words of M. Ramond, "Ce n'est assurément pas au Mont Dore que la fameuse question de la Volcanicité des basaltes sera jamais l'objet d'une discussion sérieuse!" *

The village of Murol is built upon the recent current at the base of the cone, and the space covered by the lava to the south is dotted with between thirty and forty black and shapeless eminences of scoriform basalt, which, rising from an otherwise smooth surface (for the intervening spaces have been brought into cultivation), offer a singular and striking picture. The form of the valley renders it extremely probable that, at the period of the eruption of Tartaret, there existed some stagnant body of water in this corner, which would account in a simple manner for these protuberances, it having been observed that such irregularities of surface are liable to be created whenever a current of lava meets in its progress, or rather rolls over, any marshy or

---

* Written in 1823. In the present day it seems almost incredible that up to so late a date the *igneous* origin of basalt should still have been contested by geologists of repute. But the school of Werner was even then predominant in Germany, at Edinburgh, and in several other quarters, where it was considered heresy to dispute the precipitation from some archaic ocean of all the crystalline rocks. And as it was felt that the primary crystallines could hardly be separated in the character of their origin from the trap rocks, the battle was fought about these.

moist ground: the conversion of the water into steam causing a series of violent explosions, and thereby tearing and driving upwards portions of lava, which are immediately consolidated in ragged and fantastic forms by contact with the air.

The basalt of Tartaret is compact, of a harsh texture, a dark colour, and contains crystals of augite, olivine, and numerous minute laminæ of felspar. The adjoining conical eminence upon which rises the ruined castle of Murol is of ancient prismatic basalt, a detached segment of one of the currents which skirt the valley on each side.

The Puy d'Eraignes, seen beyond in our sketch, is of the same nature; both owe their insulated position to the erosion of confluent mountain torrents.

Two recent cones, called Montchal and Mont Sineire, occur next in succession towards the south; they have exploded through repeated beds of basalt, and probably trachyte, at a considerable elevation on the flank of the Mont Dore.

Both have large and regular craters, and both have produced abundant currents of lava, which enter the valleys of Besse and Compains, and thread them for an extent of some miles, resting indifferently on former beds of basalt or on granite. The furious torrents of both gorges have in turn eaten out a fresh channel from the recent basalt to the depth often of 30 and 40 feet. The current proceeding from Mont Sineire in particular is of prodigious bulk and extent.

One remarkable and peculiar circumstance attends these cones; viz. the existence of a deep, large, and nearly circular hollow immediately at the foot of each. The bottom is covered with water, and they bear the names of the lakes Pavin and Mont Sineire; both are bordered by nearly perpendicular rocks of ancient basalt. Their position announces them to be contemporary with the eruptions of the neighbouring cones; and it seems probable that, like the Gour de Tazana already described amongst the Monts Dôme, they owe their formation to some extremely rapid and violent explosion.

In the same neighbourhood are several other minor cones of scoriæ, and four more crater-lakes, named Chauvet, Beurdouze, Champedaze, and La Godivel; which are probably owing to a similar cause, and date from about the same recent æra of subterranean activity.

A few other recent eruptions have taken place to the east of the same meridian; viz. in the neighbourhood of Coteuge, Rantieres, Brion, &c.; presenting the usual characteristic products already described on so many other points, the repetition of which may be as well omitted.

They are confined to the south-eastern slope of the Mont Dore, and none are to be found within the boundaries of the Cantal, which is out of the direction of the zone along which the most recently active vents are alone distributed.

CHAPTER VII.

REGION III.—CANTAL.

§ 1.

In original constitution and form the immense volcano whose remains occupy nearly the whole extent of the present department of the Cantal must have very closely resembled that of the Mont Dore; and the two groups, as they now exist, differ little in the nature of their erupted materials.

In figure the Mont Dore was described as resembling an irregular depressed cone: the Cantal approaches still nearer to it, since its sides slope more uniformly from the central heights. The chief circumstance to which the irregularities in the outline of the former mountain are obviously attributable is the excessive bulk of the hummocks of trachyte which the volcano has pushed forth in three or four directions. The trachytic lavas of the Cantal, on the contrary, perhaps in consequence of a superior degree of fluidity, have reached to a considerable distance without accumulating into such enormous masses, and have been subsequently covered by repeated and widely spread sheets of basalt, which give to the skirts of the mountain a more regular and gradually sloping surface.

The valleys by which this surface is intersected stretch out like rays on every side from the central heights into the surrounding country; they are generally deep, and bounded by steep and rocky walls exhibiting on each side corresponding sections of the different volcanic beds through which the excavation has been effected, and, towards their termination, cutting into the primary base by which these are supported.

Such are the great western valleys of the Goul, Cer, Marone, Mars, and Rue, which pour their waters into the Dordogne or Lot; and that of the Alagnon on the east, the only stream which the Cantal contributes to the Allier. We have already described the calcareous freshwater formation which within a limited space is interposed between the primary crystalline rocks and volcanic superstructure.

The lowest bed of this last formation generally consists of conglomerate. It is seen at the bottom and sides of every deep valley, often constituting a vast and imposing range of turreted cliffs, which, in consequence of a rude columnar division on a large scale, assume various fantastic forms. The bulk and extent of this bed are alike surprising. It is found in some valleys, that of the Cer for instance, continued with a constant thickness of several hundred feet from the base of the central summits to the distance of more than twenty miles; preserving everywhere a gradually diminishing inclination parallel to the sides of the mountain. It universally envelops or is accompanied by currents of basalt and trachyte, which show themselves at intervals without any regularity, and are often intimately blended with the neighbouring conglomerate. The constitution of this rock is more varied and disorderly here even than at the Mont Dore. In one spot it consists of a loose arenaceous tuff enclosing bouldered blocks of trachyte, basalt, and granite; in the next, of a consolidated ferruginous breccia of angular fragments of basalt and scoriæ, sometimes united by a cement of basalt itself, sometimes, within the limits of the freshwater basin, by a calcareous or argillaceous one. Occasionally the most decided indurated tuff containing angular fragments graduates into a compact rock enclosing small crystals of glassy felspar and augite, having a tendency to a prismatic rhomboidal divisionary structure, and stained throughout by the oxide of iron in stripes of a brown colour exactly imitating in their arrangement the undulating and parallel zones of a fir-plank.* On other points all fragments are absent, and the tuff is indurated, obviously by water, into a laminated shale, full of impressions of leaves and branches of trees, and sometimes passing into an earthy lignite, which is used by the peasants as fuel.

These different modifications of composition, as well as the position and appearance of the great beds of conglomerate, are reconcileable with no other mode of production than that to which was attributed the same order of rocks in the Mont Dore. Their formation was obviously contemporary with that of the lava-currents they envelop; but their nature and extent preclude all idea that they are owing solely to the projections of the volcano; while the great elevation which they attain round the central heights refutes the supposition of their being merely alluvial deposits from any body of water which may have existed at the foot of the mountain during its eruptions. It only remains then to conclude them the result, for the most part, of torrents of water tumultuously descending the sides of the volcano at the periods of eruption, and bearing down immense volumes of its fragmentary ejections, in company with its lava-streams.

These vast beds are usually surmounted by currents of basalt; sometimes, as in the valleys of St. Paul de Salers and Falgoux, repeated five or six times, and separated from each other only by a layer of their own scoriæ. I noticed no instance of beds of trachyte alternating with basalt, as was observed in the Mont Dore. Throughout the Cantal the production of the former

---

* Analogous to the regenerated trachyte observed in Hungary by M. Beudant.

rock, with its associated tufas and breccias, seems to have ceased before the eruptions of basalt commenced.

Such, at least, is the general order observed in the sections afforded by the principal valleys, at some distance from the summit of the mountain. On a further ascent, the confusion of products becomes greater, indicating the vicinity of the vent of eruption. The conical form of the whole mountain, and the divergence of all its currents from the neighbourhood of a few central heights, leads, as in the case of the Mont Dore, to the presumption that the volcano of the Cantal had one principal and central crater; and many circumstances unite to fix the situation of this upon the double basin in which the upper sources of the rivers Jourdanne and Cer are first collected.

It is from the outer circuit of this area, encircled by several culminating peaks and ridges of trachyte, that the chief volcanic currents appear to originate. In the centre rises the Puy Griou, separated from a mass of clinkstone, the Plomb du Cantal, by a considerable depression, across which passed the old road from Murat to Aurillac, now superseded by a tunnel bored through the connecting ridge at a level 900 feet lower. The Plomb, the highest point of the whole mountain (6258 feet A. E.), is of basalt, and thence probably proceeded the enormous basaltic currents which have flowed towards the south-east.

In piercing the tunnel mentioned above, a great number of more or less vertical dykes were met with, of trachyte, clinkstone, and basalt,* traversing a breccia containing cellular and scoriform fragments of these rocks, as well as veins of green pitchstone porphyry. This structure, so similar to that of the centre of the Mont Dore, is just what we should expect to find in the eruptive chimney of a volcano.†

---

* See a paper by M. Ruelle, Bulletin xiv. p. 106.
† See note to p. 127.

The eminence called the Col de Cabre overlooks the source of the Jourdanne on the north, giving rise to a mountainous embranchment principally consisting of trachyte, which extends in a north-easterly direction, and separates the valleys of Murat and Dienne; while to the N.W. of the same area the Puy Mari stands at the head of powerful and repeated currents of basalt, which, accumulating on one another, form the Montagnes de Salers, and spread from thence over a high plain towards the Mont Dore.

Indeed, with the exception of the masses already designated, and a vast but considerably degraded current of the same nature which terminates in an enormous and elevated plateau above the town of Bort on the west of the Dordogne, resting on a layer of pebbles, and apparently occupying the former bed of the river which now flows more than 1000 feet below it;—with these exceptions, the trachytic lavas of the Cantal are far from conspicuous on the exterior of the mountain, and are greatly exceeded in number, volume, and extent by its beds of basalt, which stretch in all directions from the central eminences over its sloping skirts, and on the east and south-west reach to distances of 25 and 30 miles.

To the south-east they form an extensive and uniform high plain, called La Planeze, reaching to the base of the primitive range of La Margeride, but little furrowed by watercourses, and of a singularly dreary aspect from its total nudity. At its extremity is seated the departmental capital St. Flour; and here, as wherever else this plateau is broken through, successive and parallel beds of basalt may be seen surmounting one another in very regularly columnar ranges.

A series of similar plateaux extends from the mountains behind Salers to the town of Mauriac, and thence some way northward. They consist principally of a light-coloured basalt, with a highly crystalline grain, sprinkled with large cellular cavities. It exhibits on different points the columnar, the tabular, and spheroidal concretionary structures; and wherever it has yielded to the erosion of torrents, is seen to rest on a tuff conglomerate.

But the basaltic beds through which the Alagnon has excavated its valley in the immediate neighbourhood of Murat are the most remarkable of all for their regular columnar configuration, no less than for their bulk.

They are associated with trachyte, accompanied and in part enveloped by accumulations of breccia; but on particular points colossal portions of basalt have been stripped of these coverings and isolated from the remainder of the current to which they belong. Such are the Montagnes de Bonnevie and Chastel. The former, at whose foot stands the town of Murat, has been long celebrated for the beauty of its columns. It is a large conical rock, about 400 feet in height from its base, composed of a single enormous bundle of prisms converging from all sides towards the apex; those of the exterior being slightly curved, the central ones straight and vertical. These last are the most perfect, and have been exposed by dilapidation on the eastern side of the rock. They are smooth, long, and slender, usually six-sided, rarely exceeding 8 or 10 inches in diameter, with a height often of 50 or 60 feet unbroken by joints or flaws. The museums of Paris and Lyons, as well as many private cabinets, have been enriched with columns extracted from hence; but it is a work of extreme delicacy to separate them whole from the rock, and still more difficult to convey them uninjured to any distance. This basalt is brittle, sonorous, hard, compact, fine-grained, of a dark colour, and free from any visible crystals. It is remarkable that the western face of the rock is entirely amorphous.*

---

* See Plate X.

a Fort Adrians. *b* Temple Buins, and Basaltic Rocks at Cyrene.

MOUNTAIN OF BASALTIC ACTIONS OF CYRENE, AND THE RUINS OF NEAR CYRENE.

Plate X.

29 AP 58

On the contrary side of Murat is a segment apparently of the same bed, in which the columns are subdivided by frequent joints; the separate articulations fitting into each other by means of alternately concave and convex bases, having occasionally small wedge-shaped processes arising from the odd angles of one prism, and closing over corresponding oblique truncations in the other.

Both above and below Murat, in the valley of the Alagnon, patches of the lacustrine limestone are quarried for use beneath the volcanic rocks, proving the freshwater formation to have extended to the east of the central vent of the volcano. But its chief bulk is seen on the opposite or western flank, where in the vicinity of Aurillac * a thick bed of breccia, alternately supported, covered, and penetrated by currents of basalt and trachyte, rests upon the siliceous marls and limestones, the strata of which are in many spots twisted and dislocated, as if they were still unconsolidated at the moment of their invasion by the volcanic torrents. Occasionally there are appearances of a subsequent deposition of limestone in confused strata moulded upon the rude surfaces of the volcanic beds. This may be observed between Polminhac and Yolet in the valley of the Cèr. In general, however, the line of junction of the lacustrine formation and superimposed volcanic products is better defined; it may be followed almost uninterruptedly from Vic en Carladèz to Aurillac. The currents both of basalt and trachyte occasionally send out thick vertical veins or dykes into the limestone below; but neither along the walls of these, where the contact of the two substances

---

* It is said that a sufficient quantity of gold-dust was less than a century ago contained in the sand of the river Jourdanne to repay the labour of sifting and washing; and tradition asserts that the town of Aurillac situated on this river derives from hence its name: qu. *Auri lacus?* The trachytes of Mexico are auriferous; there is nothing therefore improbable in the idea that those of France may be so likewise.—See Brieude, *Top. Médicale de l'Auvergne,* 1782-83.

is immediate, nor elsewhere, could I trace any material alteration in the texture of the limestone. In frequent instances calcareous masses, as well as smaller fragments and even some freshwater shells, may be seen enveloped by the lava, but they still effervesce briskly with acids, and, except in a somewhat dusky tinge, seldom appear at all affected.

The lacustrine beds are found at a higher level, by several hundred feet, on the east side of the mountain, than on the west, which led Mr. Raulin* to suppose the existence of a fault traversing the central heights of the mountain, and dating its formation from the first outbreak of the volcano—a supposition by no means improbable.

Without the limits of the tertiary formation, the primary base emerges from beneath the volcanic products on every side of the Cantal. The most elevated point at which I have observed it is near the village of Thièzac in the valley of the Cèr, between five and six miles from the supposed crater, where a rock of gneiss, certainly *in situ*, pierces through the calcareous strata as well as the overlying trachytic breccia, and disappears again immediately.

We have no data by which to determine the relative ages of the volcanic remains of the Mont Dore and Cantal. Appearances would lead to the conjecture that these volcanos were occasionally in action at the same period.

§ 2. Canton d'Aubrac.

The small mountainous district of Aubrac, which lies between the three towns of La Guiolle, St. Geneis, and St. Arcize, in the department of Aveyron, is for the most part covered by massive beds of basalt, which belong, I believe, to a centre of eruption

---

* Bulletin xiv., p. 174.

independent of that of the Cantal; but having had no opportunity of observing this group otherwise than from the summit of the latter mountain, I am unacquainted with the true limits and disposition of its volcanic productions. I am, however, assured by M. Le Coq, who has visited it, that the basalt has been evidently erupted from the gneiss and mica-schist on the spot, and that no traces of any tertiary or other calcareous strata are to be seen there.

CHAPTER VIII.

REGION IV.—DEPARTMENTS OF THE HAUTE LOIRE AND ARDÈCHE.

§ 1.

THE basaltic ramifications of the Cantal seem to have been impeded in their progress to the east and south-east by the high primary range of La Margeride, although some detached portions are found upon its surface. To the north of this range they extend nearly to the banks of the Allier in the direction of Lempde and Brioude. On crossing the river * and penetrating further into the department of the Haute Loire by the high road to its capital, Le Puy en Velay, we first traverse a barren tract of granitic soil, appearing to have been much devastated by torrents descending from the heights of La Chaise Dieu, and

---

[* However foreign to my subject, I must be allowed to mention, as an object of curiosity but little known, from this road being unfrequented by curious travellers, the bridge over the Allier at Vielle Brioude, one of the noblest and most beautiful works of the sort in Europe. It consists of a single stone arch, with a span of 192 feet, and rising 90 from the river. It is perfectly level at top, and, though narrow, admits of the largest carriages. The piers that sustain this magnificent arch abut on either side against rocks of mica-schist which here encase the Allier for some distance. It was built as early as the 15th century; but it is only within a few years and under Napoleon that it has been widened to admit vehicles of all sizes. The airy lightness and elegance of the arch as seen from below are not perhaps to be paralleled by any other similar piece of masonry.] In the year after this note was penned (1824), this bridge fell into the river, being probably weakened at the time the addition above referred to was made. Another equally handsome, and it is to be hoped more durable one, has since been built between the original piers, and of the same dimensions. The Allier, taking its rise in the highest range of the Lozère, is liable to floods of extreme violence. One of the most terrific ever recorded has just occurred (October, 1857), by which the works of the railway bridge, nearly completed at this spot, have been swept away, and a vast amount of other damage done.

immediately find ourselves surrounded by volcanic remains of a different character to those we left behind on the other side of the river, and evidently foreign to that focus of eruption. The Allier may, therefore, in a general way, be established as the natural boundary of the volcanic districts of the Cantal and Haute Loire.

The volcanic remains of the departments of Haute Loire and Ardèche, or, in other words, of the ci-devant provinces the Velay and Vivarais, belong to both the classes which have been distinguished above, and accordingly arrange themselves under these separate heads:—

1. The Mont Mezen and its dependencies.

2. The products of more recent eruptions, which have burst out on numerous points irregularly scattered over a broad band of the primary plateau, from Paulhaguet and Alègre to Pradelle and Aubenas, seeming to be a prolongation in a S.E. direction of the chain of recent volcanos already described in the department of the Puy de Dôme.

§ 2. The Mont Mezen and its Dependencies.

The Mont Mezen is the most elevated of an extensive system of volcanic rocks, resting partly on granite or gneiss, and in part on the Jurassic formation, which by their position and constitution prove themselves to be the remains of a single and powerful volcano, of the same character as those which have been already described in the Mont Dore and Cantal. Its products, however, are disposed in a somewhat different manner, being spread over an almost equally extensive surface without accumulating into such mountainous masses around their centre of eruption. Two causes seem to have contributed to occasion this diversity of aspect, namely: first, that the eruptions of this volcano appear to have been less frequent than in the other instances; secondly, that its lavas consist either of basalt or clinkstone almost exclusively, very little granular or common trachyte occurring among them. They, therefore, were probably possessed of great comparative fluidity; and having burst out on one of the highest eminences of the primary platform, which afforded a considerable slope in most directions, they appear to have flowed to great distances immediately upon their protrusion from the volcanic vent.

Owing to these circumstances the fundamental granite is disclosed occasionally in ravines up to the foot of the central summits; and the highest of these, the Mont Mezen, though raised on a much more elevated basis than either the Cantal or Mont Dore, is inferior to both in absolute height. It measures from the sea, according to Cordier, 5974 feet.

The Mezen itself, and the other principal masses grouped around it, are almost uniformly composed of clinkstone (phonolite, scaly or schistose trachyte), a rock which ranks first amongst the products of this volcano; while the common or massive variety of trachyte is, if not totally absent, rarely to be met with throughout the whole system. Basalt occurs in great abundance, and is associated with vast beds of its peculiar conglomerates, analogous in their general characters to the same formation in the Mont Dore and Cantal, but, owing to the non-production of trachytic lavas, with little if any pumice in their composition, and consequently almost always in the state of basaltic breccia (basalt tuff).

We shall be fully justified, by the universal declination of these different volcanic beds from the Mont Mezen, in fixing the site of their eruption in its immediate proximity; and on the south-east of this rocky eminence, in the vicinity of the Croix des Bouttières, there still exists a semicircular basin whose steep sides are entirely formed of scoriæ and loose masses of very cellular and reddish-coloured clinkstone, and which probably therefore formed a part of the circuit of one of the last-formed craters. From this spot two principal embranchments of clinkstone are projected, one to the south, the other to the north-north-west.

The first shows itself in between twenty and thirty neighbouring rocky eminences of very considerable magnitude, and more or less degraded to a conoidal form, studding irregularly the high platform of the Haut Vivarais. From the foot of one of these, called Le Gerbier des Jones, gushes the source of the river Loire.

The second constitutes a mountainous chain of numerous similar domes or cones, connected in general by their bases, and covering a wide band of country as far as the towns of Roche en Regnier and Beauzac on the northern side of the Loire. The uniformly progressive declination of this series of phonolitic summits from the Mezen to the bed of the river, where they terminate, the two last, called Miaune and Gerbison, leaning against the foot of the primitive range of La Chaise Dieu on the *opposite* bank, leads to the impression that they are the remaining portions of a single enormous lava-current, prior in date to the excavation of the actual channel of the Loire, and far the most considerable in bulk and extent of any in the Phlegræan fields of France. The space it appears to have covered is more than twenty-six miles in length, with an average breadth of six, containing therefore a superficies of 156 square miles. Its thickness must have originally been prodigious, and may be judged of by the mountainous portions still remaining, whose outline is to be seen in the accompanying panoramic sketch.* Many of these rise to a height of 400 and 500 feet from their base; and none, I be-

---

* See Plate XI.

lieve, show any marks of a division into separate beds; so that the whole colossal range must be supposed one current, the product of a single eruption. It rests generally on granite, either immediately, or with the intervention of basalt and its conglomerates; but appears also to have covered a large angle of the calcareous freshwater formation which we have already described as occurring in the neighbourhood of Le Puy.

As the determination of this fact may be considered an object of some moment, I took some pains in its examination. At the Castle of Lardeyrolle, and on different points near the villages of St. Pierre Eynac and Mercœur, the clinkstone certainly rests in part on the freshwater strata. Immediately below the latter village, which with the ruins of its castle is built on a conical peak of clinkstone, the superposition is decided and immediate. It is true that the nucleus of this and the other isolated hills capped by clinkstone which I observed within the limits of the freshwater formation, is of granite; and the great inclination of the calcareous strata of Mercœur, at the point where they support the clinkstone, indicates the granite against which they abut to be at no great distance. But it is not difficult to conceive that by this subjacent primitive nucleus alone the capping of clinkstone has been preserved from destruction; nor is it strange, *à priori*, that this rock should rarely be found resting *solely* on the calcareous strata, for the reason that the latter, consisting in this quarter of very soft and friable clays or marls, may be supposed to have everywhere yielded to the erosion of the pluvial waters, to which the highly fissile structure of the superincumbent rock would give an easy access; and owing to which, the whole of the clinkstone bed which covered these soft freshwater strata has probably been undermined, precipitated, and swept away; with the exception of such portions as were accidentally based upon some granitic knoll protruding through the marls, as in the instances quoted.

In fact, the most friable of the freshwater strata, the clays and marls, have generally disappeared; all but a few scattered patches which lean against the primary edges of their basin, or those portions that were protected by a cover of basalt and its indurated breccia, which, being less permeable to water than the fissile clinkstone, have remained longer on their treacherous foundation: but even these are reduced to comparatively small segments of beds originally very extensive, and annually suffer more or less diminution by the sapping and mining of their base.

Again, that the production of these volcanic rocks occurred subsequently to the deposition of probably the entire freshwater formation, is confirmed by a negative proof of the strongest nature, viz. the total absence of clinkstone fragments among its transported materials, or the sands and sandstones that underlie it; while, on the contrary, they abound in the alluvia and volcanic beds which were deposited upon these after they had been considerably degraded. It is scarcely possible that the bed of clinkstone could have been worn down to the insulated peaks now existing, before or during the deposition of the sands and marls, without a fragment or a pebble of it being found in these strata to attest its destruction.

Many geologists have remarked the tendency of phonolitic mountains to waste into detached masses of a conical form. Nowhere could this observation be better appreciated than along the range I am now describing, which is entirely reduced to a series of rocky eminences, presenting every intermediate gradation of figure from the rude segment of a bulky hummock to a perfect cone.

The cause of this uniformity clearly lies in the much greater facility with which this rock yields to meteoric influence on some points than on others, as well from its frequent differences of texture, and consequent aptitude to decomposition, as from its accidental varieties of structure; the columnar and laminar modifications at times combining to hasten a disunion of parts (as was remarked in the Roche Tuilière, Mont Dore), at others to afford the utmost power of resistance, as when a sheaf of columns leaning against one another converge into a pyramidal cluster. The same causes continue to influence the aspect of the mass after it has been completely isolated and reduced to a rounded form by the wasting of its angular projections. Where the phonolite is of a quality that readily decays on exposure, it presents a smooth-sided cone, clothed with a thick layer of white earthy soil, which frequently supports luxuriant forests of oak and fir. Where the rock is less destructible, its upper outline is cap-shaped, notched, and craggy, and its base encumbered with barren and ruinous piles of slaty fragments. The clinkstone of both the northern and southern embranchments from the Mezen is always, I believe, more or less laminar, often columnar. The laminæ vary from massive and compact plates of great size and thickness, to thin slates; the latter are in general use throughout the country for roofing. Its colour varies from dark bluish-grey to greyish-green, green, pale or ochrey yellow, and pure white. The lighter-tinted varieties generally occur at the bottom of the bed, never, I believe, in the upper and more exposed parts; they must not therefore be confounded with the ordinary effects of decomposition. The dark-coloured species are compact, hard, and translucent at the edges, their fracture straight and splintery; the pale parts are more granular, earthy, and approaching to common trachyte, not unfrequently of a very scaly texture, and a mild silvery or pearly lustre. Frequently globular concretions of a dark greenish hue appear thickly disseminated in a base of a lighter and greyish tint. These are evidently owing to the imperfect aggregation of augitic molecules, which under more favourable circumstances would have united into distinct crystals.

Small imbedded crystals of felspar are frequent, of augite or hornblende rarer. The lower parts of the bed are somewhat porous and occasionally vesicular: in the vicinity of the central summits highly cellular and even scoriform masses are abundant; they are usually of a reddish or reddish-blue colour, and a crystalline texture.*

BASALTS AND BRECCIAS.

I have already noticed basalt as having been largely produced by this volcanic system. The currents principally flowed towards the north-west, west, north-east, and south-east, led by the superior inclination of the primary platform in these directions.

Towards the south-east in particular an immense embranchment, composed of two or more successive beds of basalt accompanied by breccia, descends to a distance of more than thirty miles, rivalling in extent and volume that of clinkstone which we have described on the opposite side.

This current is not less interesting from its position than the magnitude of its dimensions. It appears to take its rise from between the scattered peaks of clinkstone which have been noticed above as grouped together on the south of the Mezen, and is prolonged with a gentle slope towards the south-east for

---

* I am aware that, in attributing the chain of clinkstone hills which stretch from the Mezen to Miaune to a single colossal current worn by the elements into fragments, I am differing from the greater number of local geologists, who consider these domes or puys to be each the product of a separate local outburst. I can only say that, on my recent visit to the district, all the examination I was able to make confirmed me in my original opinion. And, moreover, I felt a strong suspicion that on many points of this range the clinkstone hummocks rest upon basalt which had previously flowed as lava in that direction over the granitic surface. I would strongly recommend the investigation of these two questions to such geologists as may visit this country.

the space of about twelve miles, resting on the high primitive platform of the Haut Vivarais, and forming an elevated ridge which separates the waters of the rivers Ardèche and Erieux. Near the line of junction of the primary and secondary formations it is broken off by a deep gap, through which passes the road from Aubenas to Privas. But on the opposite side of this depression the same bed is repeated at a precisely corresponding height on the summit of a mountain of Jura limestone. From hence it is continued with a similar gradual inclination to a direct distance of about twelve more miles. As long as this bed is based on granite its breadth is trifling, and it shows itself rather as a continuous flat-topped mountain-crest than in the usual form of a wide plateau. But on entering the limits of the secondary formation it assumes a different disposition, spreads itself to a width of five, seven, and nine miles, and covers an extensive and elevated table-land, which under the ancient régime went by the name of the Coiron. There cannot be the least doubt but that the whole of this lofty tract of Jurassic strata has been solely preserved from destruction by its volcanic capping. The remainder of the formation around has been eaten into in all directions by various mountain torrents, and gnawed down by meteoric abrasion to a far lower level. The Coiron alone beneath the shelter of its basaltic coating has effectually resisted, and juts out like a huge flattened headland from the margin of the high primary plateau into the southern plain. It has not, however, escaped uninjured. The agents of waste, to which the iron hardness of basalt itself ultimately yields, have intersected it by numerous transverse ravines, excavated sometimes to a considerable depth through both the volcanic and calcareous strata. They are separated by massive parallel embranchments, which derive from a straight longitudinal axis, exactly in the manner of ribs from the spine. On either side there are as

29 AP 58

BASALTIC PLATEAUX OF THE COIRON (ARDÈCHE), FROM THE SOUTH

Plate XII.

many as eight or nine of these, each crowned by a prodigious tabular load of volcanic products, presenting in its section a vast range of vertical cliffs resting on the secondary limestone strata. The perfect correspondence in position, structure, and dimensions, of these cappings, as well as their all branching off from the same stem, testify to their having been once united in a single continuous platform; and in all their aspects, but particularly from the side of Villeneuve, where the extremities of six or seven of these ramifications may be taken in at once by the eye,* their appearance is striking in the extreme.

The thickness of the volcanic mass is usually between 300 and 400 feet, which appeared to be made up, wherever I examined it, of two enormous distinct beds of basalt, separated by a layer, varying greatly in thickness, of scoriæ and volcanic fragments united into a breccia, or of loose scoriæ alone. Each bed, but particularly the inferior one, presents a sort of lower story of very perfect and well-matched vertical columns, surmounted by a still thicker mass, which is comparatively amorphous, but on a near approach is seen to consist of innumerable small columns both straight and curved, disposed in every possible direction and entangled into every variety of figure. These two portions are blended at their line of contact in such a manner that it is impossible to doubt their belonging to the same bed. The basalt of the uppermost surface is porous, slaggy, and scoriform.†

The almost architectural symmetry resulting on many points from this arrangement was probably the origin of the fable current among the peasants of the plain below, who still call

---

* See Plate XII. Also the General Map of Central France.

† Exactly resembling in these respects the very recent lava-streams of the Vivarais to be described in a subsequent page.

these rocks "Les Palais du Roi," and imagine them to have been reared and inhabited by some giant monarch.

The limestone which supports them, corresponding in age to the Oxford clay, or middle oolite, is in parts clayey and friable, passing into and alternating with shale, and consequently yields easily to abrasion. The superincumbent cliffs are thus frequently undermined, and huge portions, detaching themselves, roll or slide down the shelving and slippery sides of the calcareous bases, which are in some parts thickly strewed with their ruins. Many of these precipitated masses consist of parcels of very regular prisms of large dimensions.

The remarkable rock on which stands the castle of Rochemaure on the banks of the Rhone, as well as two others in its vicinity, are evidently the extreme segments of one of those arms which the principal basaltic trunk of the Coiron stretches out towards the east; they appeared, however, to me not to be in their natural position, and perhaps descended by a slip or subsidence from the cliffs above during the excavation of this part of the valley of the Rhone.

Not far from the extremity of one of the lateral embranchments on the opposite side, at Villeneuve de Berg, may be observed an extensive dyke of basalt traversing in a direction nearly north and south the strata of limestone which compose some low hills at the foot of the Coiron. The basalt is both cellular and compact, very dark coloured, dense, and crystalline. The compact parts sometimes affect a globular form, at others break into rhomboidal pieces. The cavities of the cellular parts are often filled with calcareous infiltrations. The limestone shows few marks of alteration by contact with the basalt, nor do its strata appear disturbed or shifted; specimens may be broken off presenting both substances firmly united. There seems every reason to believe that this vein is not a dyke of eruption, but proceeded from an overlying mass of basalt now destroyed, the prolongation of that above St. Jean. The enormous fragments which still strew the valleys of the Laduegne and Escoutay are sufficient to prove that the bed did not originally terminate at its present limits in that direction.

The ravine called Les Balmes de Montbrul, which M. Faujas considered to be a volcanic crater, is merely an accidental excavation, I believe, in the basalt and scoriaceous breccias of this bed.

On the whole there are many circumstances, besides their immense bulk, attending the basaltic currents which cap the Coiron, that render them in a geological light perhaps the most instructive of any in the interior of France. Their disposition at first in a narrow ridge across the granitic heights, then in a widely-spreading sheet over the secondary formation, perfectly parallel to its strata, and everywhere preserving the same gradual declination from the neighbourhood of the Mezen to the extremity of the Coiron, evidently demonstrates their having flowed at an epoch when the surface of the primary platform was directly continued in that of the secondary strata; and appears to indicate that this last formation at least was at the time in a state of comparative integrity.

That it had, however, long emerged from the ocean in which it was deposited, is attested by the circumstance that beneath the basalt a vegetable soil is found containing terrestrial shells of a species still existing in the same country (cyclostoma elegans).

The immense quantity of matter which must have been abstracted from the secondary district of the Rhône valley since the epoch at which this lava was emitted—an epoch which this last-mentioned fact proves to be but recent (geologically speaking)—cannot but strike us with astonishment. There can be no doubt that the surface on which the basalt of the Coiron rests was at that period the lowest of the neighbouring levels, or these repeated currents of liquid matter could not have flowed in its direction; yet at present this same surface vastly overtops every other height of the same formation, and ranges upwards of a thousand feet above the average level of the valley-basins of the Ardèche and Rhône on either side. That a considerable proportion of these was excavated by "rain and rivers," in other words, by meteoric agency such as is still in operation, and not by any diluvial or general flood, is susceptible of direct proof, as will shortly appear from our examination of the volcanic formations of the Bas Vivarais. To attribute, therefore, the remainder to any other cause of an hypothetical nature unsupported by evidence, would seem to be contrary to the rules of analogy. But the conclusion that the greater portion of the valley of the Rhône has been so recently excavated, and by such agency alone, involves important consequences; since the same agents must have been at work everywhere else, and produced results as stupendous during the same (comparatively) recent period. These considerations will be resumed hereafter.

I now proceed to the other basaltic remains which owe their origin to the volcano of the Mezen. They form the covering of a very extended surface of the Haut Vivarais and Velay, but, from their situation on so elevated and exposed a region, have been intersected in all directions by the channels of mountain torrents, so as to disclose the subjacent rock, which is everywhere granite, except within the boundaries of the freshwater formation of Le Puy.

It is remarkable that detached plateaux of basalt are scattered profusely near the borders of the great phonolitic embranchments noticed above. Not unfrequently the interval between two or more conoidal rocks of clinkstone is occupied by a flat superficial bed of basalt extending quite up to their bases; and it is very difficult, in consequence of the excessive degradation of the clinkstone and the piles of débris which gather round their skirts, to ascertain whether the basalt passes below them, or has been deposited since their separation. The former conclusion is, perhaps, the most probable.

The basaltic beds which flowed eastward from the Mezen reach from Fay le Froid to beyond Saint Agrève; constituting the crest which separates the waters of the Loire from those that flow into the Rhône. On many points they have been reduced to scattered patches by the same causes which have cut up and ravined to a great depth that barren and dreary primary canton called Les Bouttières. Those which took their course to the north and north-west are more entire, descending into and traversing the valley of the Loire, accompanied by large accumulations of conglomerate. I think it probable that the basin of the freshwater formation at the period of its invasion by these volcanic products had been laid dry, and was even partially eaten into by the torrents that descended from the environing heights: that the position of the river which then effected its drainage and received these streams was much to the west of the actual channel of the Loire, and at the foot of the western granite range: and that in this state of things the whole of the basin up to that range was deluged by vast currents of basalt, which flowed from the vicinity of the Mezen in company with an immense quantity of fragmentary materials, transported by torrents of water and deposited in the same disorder as the conglomerates of the Cantal and Mont Dore.

The facts on which this opinion is built are contained in the following observations, which will be made much clearer by the accompanying panoramic sketch of this district.*

---

* See Plate XI.

From the centre of the volcanic system near the Mont Mezen vast beds of basalt are seen to stretch with a gradual inclination over a wide surface of granite, towards the west and north. On attaining the freshwater formation they are prolonged over it exactly in the same manner, without any material change in the angle of their slope and few complete interruptions to their continuity. Where these do occur, the original union of the different segments is evident from the corresponding altitude, disposition, and nature of the masses which border either side of the valley. In this way they extend to a considerable distance beyond Le Puy; and though towards the extremity of their course they have been themselves covered by more recent and scarcely less copious lava-streams from a different source, they may still be universally recognised by their peculiar characters, along the mural flanks of the numerous ravines which intersect this plain, almost up to the base of the western granite range.

While describing the freshwater formation of Le Puy, I spoke of it as arranged in strata having a progressively decreasing dip from the granite against which they lean. It is remarkable that the volcanic beds are not parallel to the planes of these strata by which they are supported, but cut across them irregularly, and often very abruptly; as for example in the Montagne de Doue, where the dip of the clays and marls is scarcely 10°, while that of the superposed bed of basalt and breccia is in some parts 40° and 50°.

Nor do the volcanic beds in general rest immediately on the calcareous strata, but with the intervention of a layer of clayey and micaceous sand mixed with volcanic ashes, and the débris of granitic and volcanic rocks, particularly rolled fragments of clinkstone. This stratum is not parallel to those of the freshwater formation on which it rests, but to the volcanic beds by which it is covered, and was evidently deposited in channels already deeply excavated through the soft calcareous strata below. It generally encloses vegetable remains, slightly carbonized, and on certain points so abundant that this alluvium passes into a lignite, of which some beds are occasionally of sufficient thickness to become the object of extraction, as at Aubepin and La-roche-lambert. The plants are either reeds and grasses, or leaves, twigs, and branches, of dicotyledonous trees. M. Aymard pronounces the opinion, that the *Flora* of this alluvium is scarcely, if at all, distinguishable from that still existing in the district. The *Fauna* of the same beds (which are very ossiferous), on the other hand, contains many species not only unknown in France, but altogether extinct; especially is this the case with the mammiferous animals, among which are found here the remains of three species of mastodon, several carnivora—machairodus, hyæna, &c.; of Pachyderms, a rhinoceros (Mesotropus, Aym.), and a tapir, &c. (See the catalogue subjoined in Appendix.)

Where this intermediate layer of alluvium is absent, and the volcanic products rest directly on the freshwater strata, the line of contact exhibits a breccia of scoriæ and puzzolana enclosed in a calcareous or argillaceous paste, such as has been remarked in parallel situations in Auvergne. In some cases this breccia appears to proceed in part from decomposition of the volcanic beds above, and would be classed mineralogically as a wacke. It is remarkable beneath the mass of basaltic breccia called Les Rochers de Peylenc, for containing numerous nodular blocks of olivine of an extraordinary size. One of an oval shape which I measured was five feet in circumference, many others around were nearly as large. These masses are rounded as if from bouldering; yet those which are embedded in the basalt above have the same figure;* they break

---

* They have, I conceive, been rounded by the internal friction of the lava which enclosed them, as it flowed onwards in a state of imperfect liquidity. See Scrope on Volcanos, p. 104.

easily into angulo-globular pieces, and are of a transparent green colour irregularly spotted with a darker green when unaffected by decomposition; those in which this process has begun are more friable and opake, and of a dull yellow or orange hue.

The same hill is interesting from the circumstance that the river Sumene, which seems once to have flowed in a different direction, has evidently eaten its present channel (and that quite recently) through the clayey marls supporting this great bed of basalt and breccia, part of which has been in consequence precipitated. The piles of rocky fragments, which the river alternately dashes over and dives between, with the formidable walls rising on each side, adorned with magnificent columnar ranges, and almost closing over the darkened gorge, form a striking scene of ruinous grandeur.

The volcanic bed itself, wherever it appears, may be considered as an extensive tabular mass of indurated breccia from 100 to 300 feet thick, in turns supporting, overlying, and enveloping currents of basalt, which at times diminish to the dimensions of a narrow vein, at others swell out into the amplitude of an enormous bellying mass. In the latter form they have been frequently stripped of the enclosing conglomerate, and appear as shapeless and almost isolated rocks.

The breccia is in some places stratified, and has often a tendency to a rude columnar configuration. It consists of both angular and rolled fragments of volcanic rocks of many kinds, of granite, gneiss, and mica-schist, and occasionally of marly limestone; the volcanic of course predominating. The cement is generally of volcanic ashes, hardened by water and pressure, participating but little in the pumiceous nature of trachytic tuff, and passing into an arenaceous compound, the detritus of the rocks whose fragments it encloses. Sometimes the gluten is merely an oxide of iron, at others basalt itself fills the interstices of the broken materials, appearing to have enveloped them in a state of fluidity.

The basalt varies in character in different parts, seeming to belong to separate currents. It is, however, generally columnar; often presenting the greatest accuracy of design, as in the rock behind the mansion of Douc, and on numerous other spots along the valleys of the Loire and its feeders. On more than one I have observed the three chief modifications of divisionary structure—the columnar, tabular, and spheroidal—distinctly exemplified in the different parts of the same mass, and passing into one another. Its mineralogical characters are extremely diversified. Sometimes as we have seen, imbedded knots of olivine predominate, sometimes crystals of augite, of hornblende, or of glassy felspar. Many varieties are much iron-shot, and then black, compact, hard, and sonorous; others have almost an earthy feel, the colour of a dead leaf spotted with bluish-grey, are porous, and decompose rapidly, separating into angular globules. Such is that which occurs on each side of a small ravine running up from the village of Expailly, near Le Puy, to the foot of a hill called Le Mont Crousteix, and which contains the far-famed zircons, sapphires, and garnets of Expailly. They have been for many centuries found in the sand of the streamlet called Rioupezzouliou, that drains this ravine; but it is only of late that they have been detected in their matrix, which is the above-described basalt. The zircons occur in small imperfect crystals, and are collected in such numbers that I have seen a large drawer full. The sapphires are next in abundance, and are found in hexahedral prisms rarely exceeding three lines in diameter. The garnets are scarce, and in rhomboidal dodecahedrons, usually about the size of a large pea. But all these crystals, not only when taken from the alluvium of the ravine, but those also which are imbedded in the compact basalt, have lost the sharpness of their angles, and are in general much defaced. There is, indeed, every reason, from their aspect and mode of setting, to suppose that their crystallization was not contemporaneous with that of the felspar, augite, and olivine which they accompany, but preceded it, and that these minerals resisted the fusion to which their original matrix was probably subjected before its conversion into a basaltic lava.

Both basalt and breccia occasionally envelop irregular deposits of volcanic ashes having the appearance of desiccated clay; and this substance, where in contact with the basalt, is divided to the depth of about a foot into very regular columnar prisms, perpendicular to the surface of the basalt with which it is in contact, and whose most general mode of structure it accurately imitates in miniature. It is often of a fine texture, and a rich salmon colour; sometimes graduates into a micaceous sand, the detritus of volcanic and primitive rocks, similar to that which occasionally forms the paste of the basaltic breccia, and on other points into a conglomerate of bouldered pebbles.

The valleys worn through this vast volcanic bed, and the lacustrine strata beneath it, offer here, as in the Cantal, an extraordinary series of pictures. The breccia, where it is much impregnated with iron, or otherwise considerably indurated, has resisted the agents of erosion, and remains in isolated masses of grotesque forms. Such is the Rocher de Polignac, that called Corneille, whose calcareous base is half encircled by the town of Le Puy, and that of St. Michel, a needle-shaped and picturesque pyramid, crowned by a church, of which a tolerable engraving is given by Faujas de St. Fond.*

---

* The tendency of coarse conglomerates to waste into pyramidal forms is generally known. Its cause is disclosed in the striking instance of the

In frequent instances veins or dykes of basalt penetrate the subjacent strata, whether these are breccia, limestone, or granite. Examples of the first kind are exceedingly common; one of the second may be observed at the foot of the hill called Brunelet, on the left of the road from Le Puy to Issingeaux. The dyke proceeds from a mass of basalt, and cuts vertically through strata of marly limestone. Its lower termination is not visible. The road from Le Puy to Rozières, on the descent towards Mercœur, is traversed by a dyke of basalt about a foot and a half wide, and which can be traced only about seven or eight yards in length. It occurs exactly where the marly strata lean against the granite, and appears parallel to the plane of their junction, which is elevated at an angle of about 60 degrees from the horizon. One wall is unaltered granite, the other a narrow stripe of granite bleached and disintegrated to the state of gravel, intervening between the basalt and the calcareous strata. The basalt is accompanied by scoriaceous portions, and from its similarity of texture, and other mineralogical characters, as well as its position, appears to have derived from a conical mass still capping the granite at a higher elevation.

The side of a narrow ravine at a spot called Les Pandréaux, between Lantriac and Le Puy, presented to view another nearly vertical basaltic dyke 3 feet wide, cutting through granite, which is considerably altered along the line of contact; and at no great distance, in the same valley, rises the Roche Rouge. This is a swelling portion, or "renflement," of a basaltic dyke, which, from its hardness of substance and solidity of structure,

---

"Pyramids of Botzen" in the Tyrol, each of which evidently owes its preservation to a large boulder which caps its apex; whence it follows that the amount of excavation produced in the conglomerate bed of which these and similar pyramidal remnants formed a part was effected solely by the force of direct vertical rain. See note to p. 106.

has resisted the abrasion that has gradually worn away the granite around (of a very friable quality), from which it now protrudes to the height of about 60 feet, with an average thickness perhaps of 30. The continuation of this dyke is visible on each side, appearing like a narrow vein often not more than a foot wide, running from the Roche Rouge towards the north-west and south-east. It may be traced to a distance of 500 or 600 yards; its course being rather sinuous than straight. The basalt is partly scoriform and cellular, partly divided into small columns laid perpendicularly to the walls of the vein. The granite of these is scarcely at all affected, and sometimes adheres firmly to the basalt.

A question may be raised on the mode of formation of this and other similar dykes, namely, whether they were propelled from above or below; in other words, whether each is the chimney of a volcanic eruption, or a portion of basalt from a current above, which in the state of lava insinuated itself into a fissure pre-existing in the granite. In support of the latter hypothesis it may be observed, in the case of the Roche Rouge, that the bed of basalt of which the Montagnes de Douc and Servissas are the principal remaining segments, certainly once extended over this spot, and that instances of such currents having pushed forth considerable dykes into the granite on which they rest are, as we have seen, by no means uncommon, while those who prefer the opposite opinion may allege the freshness of aspect, both of the scoriæ and basalt of the Roche Rouge, and the frequency of recent volcanic eruptions in the neighbourhood (to whose description we are now arriving), each of which must, in all probability, have left some such dyke as this filling the fissure through which its lava ascended. For myself I incline rather in the instance of this dyke to the idea of its ascent from below; and that we have in it an example of the condition which every ordinary volcanic vent productive of but a single minor eruption of basaltic lava would exhibit, when stripped of the outward coverings of scoriæ and other fragments that in almost all cases conceal the orifice in the superficial rocks through which these matters were protruded. There appeared to me to be a separate axis of basalt penetrating the heart of the Roche Rouge, as if the lava had continued to be expelled through a central channel, after the external parts which were in contact with the granite had been consolidated to a considerable depth. The rock of St. Michel seems to contain a similar dyke, which may probably have been erupted on this spot. It is, however, of course evident that the conglomerate of which it is composed must have been originally enveloped and supported by surrounding beds of softer materials, since worn away by aqueous erosion, and the same is true of the Rocher Corneille, that of Polignac, &c. The existing valley of the Borne, as well as that of the Loire into which it opens, must have been excavated subsequently to the production of these remarkable rocks, whatever may be their date. On this point, however, the extreme similarity of these basaltic breccias to those of the Cantal attests their parallel origin. Though therefore the dykes by which they are occasionally penetrated may be considered the result of local eruptions, I cannot doubt that the breccias (of Corneille, &c.) are derived chiefly from the Mezen.

§ 3. The Products of more recent and scattered Eruptions.

We now come to the volcanic remains of the second class occurring within the departments Haute Loire and Ardèche. These are the products of a later epoch of volcanic activity, and almost uninterruptedly cover a broad zone of the primary platform from a point north of Paulhaguet to Pradelle and Aubenas in the south. They constitute a prolongation of the chain of puys of Auvergne, but do not appear in any instance of quite so recent a date as the latest of those.

The various points on which these eruptions have broken forth are still marked by numerous volcanic cones of scoriæ, whose projection, as in Auvergne, accompanied the development of the volcanic phenomena. They are so thickly sown along the axis of the granitic range that separates the Loire and Allier from Paulhaguet to Pradelle, as generally to touch each other by their bases, and form an almost continuous chain.

On both sides of this they are more sparingly distributed, dotting here and there the slope towards either river; a few being also found on the further side of each. Between Pradelle and Aubenas they occur more rarely, and none to the south of the latter town. There is, however, a large and productive group to the north-east of Pradelles, in the vicinity of Prezailles. Throughout the above-mentioned tract I counted more than a *hundred and fifty* of these cones, and probably omitted many.

They are not altogether in the same state of freshness and preservation as the great proportion of those we have noticed in Auvergne. Not many present an entire or even a distinctly marked crater, and the generality have wasted to ridgy saddle-shaped hills, a form which volcanic cones have frequently been observed to assume by degradation. As usual, they are solely composed of slaggy and funiform scoriæ, lava-bombs, lapillo, volcanic sand, fragments of granite and basalt, and sometimes massive portions of this last rock *in situ*, the relics of the currents which descended their flanks or boiled up within their craters from the vent beneath. Their surfaces are scantily clothed with a meagre herbage, and occasionally a few stunted Scotch firs;

but their dilapidation is incessantly going forward by means of frequent and shifting surface-rents.

The lava-currents, whose emission accompanied the formation of the cones of the principal chain, must have been exceedingly abundant. According to the position of each point of eruption with respect to the slopes of the granite range, they appear to have directed themselves either to the east or west, descending into the bed of the Loire on one side and of the Allier on the other. The former have covered nearly the whole eastern slope of the range (the granite which forms its nucleus appearing only at distant intervals or in ravines worn through the basaltic beds), and are continued over the freshwater strata in a uniform sheet, forming a very extensive and but slightly inclined tract, which they seem to have completely deluged. There is every indication that, by these prodigious and perhaps nearly simultaneous irruptions of lava upon the basin of the Loire, the river was driven from its former bed and forced back upon the slope formed by the similar products of the Mont Mezen. It probably long erred between the two, every fresh invasion partially shifting and raising the level of its channel, so that its actual valley, and those of its tributaries (especially of the Borne), are excavated as well through the vast mass of earlier basalt and breccia, which we have seen chiefly proceeded from the habitual vent of the Mezen, as through an uppermost bed, generally single, of basalt alone, which, in its fresher aspect and uniform inclination from the chain of volcanic cones, is proved to derive from them. These later lavas are separated usually from one another, when repeated, and also from the older basalts and breccias when they overlie the latter, by beds of conglomerate containing scoriæ, fragments of granite, &c., and often layers of bouldered pebbles, showing the currents to have occupied, as usual, the shifting beds of the rivers and mountain torrents. Where the fragmentary matter has been taken up and deposited by these torrents, breccias have been produced identical with those of an earlier epoch.*

These changes in the position of the Loire seem to have been accompanied by the progressive excavation of that deep and narrow gully through which this river now escapes from the basin of the freshwater formation of Le Puy,† and which could not have existed at the period of the flowing of certain beds of basalt (the plateaux of Chambeyrac and l'Oulette) that are seen capping its perpendicular cliffs; or it must have been completely choked up by the lava to the level of these plateaux, like the valleys of the Bas Vivarais, which we shall shortly have to notice. This channel has been worn irregularly across the high granitic embranchment of Chaspinhac which separates the basin of Le Puy from that of Emblavès (a minor division of the same freshwater formation), and may perhaps have been partly occasioned by the creation of a fissure in that direction through some earthquake attendant on the volcanic eruptions. It is even probable that the waters of the lake were previously discharged by another outlet, perhaps over the low part of the granite ridge between Blaves and St. Etienne.

Between these two points may be seen another striking proof of the vast amount of erosion effected by the existing torrents, in the gorge of the Sumène, a stream which joins the Loire just before it enters the defile at Pierre Neyre. On the left of this deep hollow the basaltic plateau called Chauds de Fay rests upon tertiary strata. On the right, the cone of scoriæ called Mont Serre is perched upon the edge of the granitic escarpment

---

* M. Aymard (from whom I differ with great deference to his superior local knowledge) is of opinion that these basaltic beds belong to a period much earlier than the volcanic cones of scoriæ which they surround. So far as I understand his views, he appears to suppose the eruptions which threw up these numerous cones to have produced no lava-streams! This would be so contrary to all analogy, not only of the Auvergne and Vivarais north and south of the Velay chain, but also of volcanic phenomena in general, that I cannot for a moment assent to its possibility.

† See Plate XI.

of Chaspinhac. And yet the position and inclination of the basalt of Fay proves it beyond all doubt to have flowed as a lava-stream from this vent across the space now occupied by the deep channel of the Sumène, which could only have been excavated by its existing stream (aided, perhaps, by some earthquake fissure), since the entire condition of the cone of loose scoriæ above rebuts the supposition that any general denuding wave can have passed over these heights since its production.

After traversing the lower basin of Emblavès the Loire finally issues through another similar defile, where its passage seems once to have been effectually barred by the great current of clinkstone already described, of which two colossal remnants, the rocks Miaune and Gerbison, rising on either side the gorge to a height of 1800 feet above the river, attest the original elevation. That this enormous dyke suddenly thrown across the valley of the Loire must have caused its waters to accumulate into a lake covering the whole extent of the freshwater formation, is extremely probable. It is not, however, to this event we can attribute the original creation of the lake in which the freshwater strata were deposited; for the granite which supports the bases both of Gerbison and Miaune at a height of 1100 feet above the present river would alone have formed a sufficient barrier for that purpose; and we have also seen that there is reason to conclude the clinkstone more recent than the freshwater formation.

The gradual erosion then of the gorge of Chamalière, through both the mass of clinkstone and the granite on which it rested, was probably coeval with that of the channel of discharge of the upper basin; and some remnants of basalt which show themselves attached to the sides of the excavation at Pierre-neyre and Artias prove the former to have been occupied at intervals during this process by lava-currents from the neighbouring vents of eruption.

Among the puys of the Monts Dôme we are enabled by their comparatively rare occurrence, and the intervals of primary rock which separate their products, to trace almost every stream of lava to the crater which marks the spot of its emission; but in this chain of vents the cones are more numerous and closer, the volcanic energy seems to have been exerted far more furiously, and the lava-currents to have united into one continuous and enormous crust, where all are mingled and confounded.* An extensive and gently sloping plain has thus been created, appearing thinly dotted by the cones which have been thrown up to the east of the principal line. Each of these has in all probability furnished its contingent to the sea of basalt that overspreads the plain; but all attempt at appropriation is out of the question. A few show themselves in the vicinity of Le Puy, having burst through the great volcanic bed which, descending from the Mezen, had previously invaded the basin of the freshwater formation. Such are the Montagnes de Denise, Ste. Anne, Seinzelles, Crousteix, Eysenac, and Mont. In their scoriæ are implanted numerous fragments of granite and gneiss: which prove that the source of their eruptions was more deeply seated than the breccias and basalts through which they burst. To one of these eruptions, and probably to that which produced the double hill of Mont, south-east of Le Puy, is owing a small platform of

---

* I again refer to the parallel instance of the chain of volcanic cones thrown up in the island of Lancerote, one of the Canary Isles, by the tremendous volcanic eruptions to which that island was subjected between the years 1730 and 1736. (See p. 95, above.) A very interesting description of these volcanic products, accompanied by a relation of the principal occurrences of the eruption, from a MS. narrative written by an eye-witness, has been published by M. de Buch. The formation of 30 distinct cones on a fissure of great length, within so short a space of time, leads to the supposition that many of the eruptions which threw up the puys of Auvergne and the Velay were similarly circumstanced. Not that the whole chain was contemporaneously produced, for this is opposed to the numerous appearances of unequal antiquity already dwelt upon at length in their description, particularly with respect to those of the Monts Dôme; but that epochs of intense activity alternating, as is the usual character of volcanic action, with lengthened periods of quiescence, frequently gave birth to a great many eruptions from independent vents within a short space of time, creating as many distinct cones, arranged either in lines or groups, and as many streams of lava.

columnar basalt called Montredon, rising in the middle of the valley of the Borne, and resting on a shingle of primary and volcanic boulders, evidently at one time the bed of the river, though now more than 50 feet above its present channel. This seems to have been one of the most recent eruptions of the district. And it is remarkable that in no instance are lava-streams found, as in Auvergne, occupying the river-channels of the existing valleys, as if they had flowed but yesterday.

The cone that has attracted most observation, owing to its immediate vicinity to the town of Le Puy, is that of the Montagne de Denise. And it does, in fact, offer some peculiar features of an interesting and problematical character. The summit and flanks of this oblong hill are covered with large accumulations of very fresh-looking scoriæ, lapillo, and puzzolana, out of which several prominent masses of basalt are projected into the valleys around and beneath. One of these forms a bulky promontory, descending to the level of the river Borne on the south, and exhibiting two colossal ranges of basaltic columns, one above the other: the upper is called the Croix de la Paille, the lower (which hangs in a cliff over the river) the Orgues d'Expailly. At its sides and base this basaltic mass is enveloped and passes into a stratified and sometimes laminated breccia or tuff of no great coherence which clothes some of the outer slopes of the hill. Other massive rocks of a breccia or peperino similar to that of the Rocher Corneille in the town of Le Puy, constitute the nucleus of the hill itself, and through these the eruption of the more recent lavas and scoriæ evidently broke out. This massive and indurated peperino is of much earlier formation than that which accompanies the erupted lavas and scoriæ, although the two breccias can with difficulty be distinguished, and in places appear almost to graduate into each other. There would be little remarkable in this were it not that in the last-mentioned stratified deposits large quantities of bones are found of elephant, rhinoceros, Cervus elaphas, and other large mammifers, and in one locality the undoubted remains of at least two human skeletons. A block of this breccia containing the greater portion of a human skull and several other bones is preserved in the museum of Le Puy. The matrix in which these fragments are firmly embedded is unquestionably a portion of a stratum of indurated tuff or breccia which envelops and passes into the basaltic lava of Denise. The examination I made on the spot, which is just above a house called the Hermitage on the road from Le Puy to Brioude, left no possible doubt of this fact on my mind. It was discovered in 1844, and the attendant

9. Mont Denise, near Le Puy, from the South-West.

10. Mont Denise, near Le Puy, from the South-East.

1. Old Breccia Rocks of the Col.
2. Road from Le Puy to Brioude.
3. Croix de la Paille.
4. Orgues d'Expailly.
5. (In lower cut.) Spot where human bones were found in strata of tuff.
5. (In upper cut.) Spot where bones of rhinoceros, elephant, &c., abound.
6. Castle of Polignac in distance.

circumstances were carefully inquired into and reported to the Academical Society of Le Puy by M. Aymard. At the meeting of the Scientific Congress of France which took place at Le Puy in 1856, the question of the genuineness of this specimen was discussed, but no reasonable doubts could be thrown upon it, and the great majority of the savans who examined the question were of that opinion.* Nor, in truth, need there be any surprise at the discovery that this district was inhabited by man at the period when the most recent volcanos were in eruption. The more remarkable inference from the discovery of these bones is that man had for his contemporaries in this country several races of extinct mammalia, of the genera rhinoceros, elephant, &c., whose remains are found in the similar stratified breccias occupying the same position on the slopes of the same volcanic cone. It is also evident that vast changes must have taken place in the configuration of the country since the eruption of the volcanic vent of Denise, and consequently since its occupation by man. The valley of the Borne must have been greatly widened and deepened, if not entirely excavated, since that epoch, and the time required for these changes throws back the date of the eruptions which buried in its ejections the two human skeletons of Denise to a very distant period. On the other hand, it is evident to any one who examines these valleys which drain the basin of Le Puy within the limits of the freshwater beds that the progress of denudation is exceedingly rapid. The red and blue clays, gypseous marls, and flaky calcareous shales are readily undermined by the torrents which, however small in summer, in winter descend with great violence in this high region, and their débris are at once carried off in the shape of mud. Even the load of volcanic rock which is precipitated as the underlying strata give way

---

* Congrès Scientifique, &c., 1856, vol. i. p. 282.

is soon reduced to pebbles and hurried off. In the Vivarais we shall have occasion to observe this process going on very obviously, and in a manner strikingly suggestive of the prodigious results it is calculated to effect within no very great lapse of time.

In the majority of instances, as near Chaspinhac, St. Geneys, Couron, Alègre, &c., the cones rise immediately from granite. That in the neighbourhood of the last town, called *La Montagne de Bar,* has a large and regular crater at its summit, about a mile in circumference and 150 feet in depth. A flat area 700 feet in diameter occurs at the bottom, which once contained a small lake or pool of water, but has been artificially drained by a channel cut through the lowest part of the encircling ridge.

The lava-streams which descended from the principal range of cones towards the Allier have in a similar manner encrusted the western slope with a thick coating of basalt, and appear to have occupied the former bed of the river nearly in its whole extent from Langogne to Vieille Brioude. The primitive chain of La Margéride, however, rising immediately from the western banks of the Allier, drove the river back upon the basalt that had usurped its channel, and through this, as well as the granite beneath, it has excavated a fresh one. By this process have been disclosed many most magnificent mural ranges of columnar basalt which at St. Ilpize, Chiliac, St. Arçon, Monistrol, &c., encase and frown over the river. They are generally seen to rest upon a bed of water-worn pebbles, from 100 to 150 feet above the present stream, and may be traced eastwards uninterruptedly to the volcanic cones on the slope or summit of the range above.

The basalt originating in this linear group of volcanic mouths assumes on different points a very regularly columnar, a tabular, and a spheroidal concretionary structure. It sometimes also separates so readily and to such a degree into small angular globules from the size of a nut to that of a millet-seed, that the roads are strewed with them to the depth of some inches, and the foot often sinks into rocks of this nature as into a heap of gravel. In mineralogical characters the rock varies from distance to distance. Among the remarkable kinds I noticed one with very large spherical cells and an exceedingly crystalline texture; the interlaced grains being felspar, augite, and a bright yellow transparent olivine. It occurs round La Roche on the road from Le Puy to St. Privat; another, near St. George d'Aurat, dense, heavy, and iron-shot, contains still larger oblong cavities frequently coated internally with fiorite.

In this range also, as in the Monts Dôme and Mont Dore, we meet with a few lakes occupying wide, deep, and nearly circular basins, which bear every appearance of having resulted from some violent volcanic explosions, but differ from ordinary craters, not only in their greater dimensions, but in the nature also and disposition of their enclosure, which is usually of primary, or, at all events, pre-existing rocks, merely sprinkled more or less copiously with scoriæ and puzzolana, little if at all elevated above the surface of the environing country. Such are the lakes du Bouchet, de Limandre, d'Issarles, de St. Front, as well as a large and remarkable hollow, now drained, in which the river Fontaulier takes its rise, and which is traversed by the road from Usclades to Montpézat. The latter crater contains a small parasitic cone rising from its bottom. I need not repeat here the remarks upon the peculiar modification of the volcanic phenomena to which this variety of crater apparently owes its formation, which were sufficiently dwelt on in the description of the Gour de Tazana.*

Between Pradelle and Aubenas the cones diminish in number, rising here and there through the great forest of Bauzon, and showing themselves up to the escarpment of the elevated platform of the Haut Vivarais.

---

* P. 81.

But the most remarkable and interesting by far of the recent volcanic remains of the zone we are now considering, and perhaps of all France, are those that occur on the steep declivity by which this escarpment is connected with the great southern valley or low-lands of the Bas Vivarais and Languedoc.

I have already described the primary table-land as abruptly cut down on this side. Its rapid slope is intersected by deep mountain gorges, into which frequently open the transverse trough-shaped valleys of the coal formation. Viewed from below, this front of the great platform appears as a precipitous curtain-like range, broken by recesses into short, steep, and massive promontories, in which all the rude and stupendous scenery of a granitic mountain district is displayed in its full magnificence.* It is therefore an unexpected and striking contrast that is presented by a few single and regular volcanic cones perched at distant intervals upon the rocky ridges of these granitic embranchments; nor is it less surprising to find the gorges that sever them almost choked up to some distance by enormous flat-topped beds of columnar basalt, such as we have been accustomed only to observe as the cappings of elevated hills.

These remarkable characters belong to six different points of eruption, designated by the six very perfect volcanic cones of Montpézat, Burzet, Thueyts, Jaujac, Souillols, and Ayzac.

1. The first, called by the peasants *La Gravenne de Montpézat*, from the puzzolana or "gravier" of which it chiefly consists, is a

---

* It would be perhaps difficult to find, in any range of mountains, scenes which present a more exquisite combination of beauty and magnificence than some of the valleys of the Bas Vivarais, so little visited by hunters of the picturesque. The rich glow of their chesnut forests, tinted by a soft and brilliant atmosphere, is far more adapted to painting than the cold transparent colouring of the Alps and Pyrenees, their pine-forests and waterfalls; nor can the outline of their masses be considered as much inferior in grandeur. The scenery is in fact that of the Apennines, but with a more luxuriant vegetation than that great *limestone* range can support.

29 AP 58

Plate XIV.

d. Volcanic Cone, called Gravenne de Montpezat.
c. Gneissic Height of Haut Vivarais.
b. Castle of Pourchirol, on Basaltic Plateau.
a. Another Volcanic Cone surrounded by Granitic Heights.
Confluence of Rivers Fontaulier and Pourseille.

VALLEY OF MONTPEZAT (ARDÈCHE).

cone of very large size formed on a granitic ridge which separates the rivers Fontaulier and Ardèche. It has a very regular bowl-shaped crater, gently inclined to the north; in this direction the lava poured over its lowest lip into the basin of Montpézat, which it entirely filled to an average depth perhaps of 150 feet, and a width of nearly half a mile. Hence it descends the valley to the confluence of the Fontaulier and Burzet, where it appears either to have stopped, the volcano having exhausted its efforts, or to have mingled with a similar current which reached the same point from above the village of Burzet. The mass of basalt thus deposited, as well as the subjacent granite, has been since cut through to depths varying from 100 to 200 feet, by the powerful action of the rivers whose channels it usurped; its present disposition, and the beautiful columnar ranges discovered by this excavation, may be imperfectly judged of from the annexed sketch taken near the junction of the torrents Fontaulier and Pourseille, about a mile below Montpézat.* Similar ranges extend all the way to its termination at Aulière, where they are still more symmetrical, many columns being straight, vertical, and entire from top to bottom of the escarpment. The engraving published by Faujas, if its faulty execution be allowed for, will give a tolerably correct idea of the architectural regularity of this façade. The basalt may be observed on many points to rest on granite, with the intervention of a stratum of rolled pebbles. Its uppermost surface is bristled with rocky and scoriform projections, which, however, by decomposition resolve themselves into a rich soil affording nourishment to very productive chesnut forests.

Some way above Montpézat, and perched just below the verge of the high primitive platform, is another cone of puzzolana and

---

* See Plate XIV.

scoriæ, also seen in the engraving, which does not appear to have given birth to any lava-stream.

2. *Volcano of Burzet.*—A bed of basalt occupies the bottom of the valley of Burzet, and follows all its windings as far as the point where it opens into the Ardèche, a distance of eight miles. It was produced from a point of eruption considerably above the village of Burzet, and at about the same elevation as the last-mentioned cone. It is chiefly remarkable for imbedding numerous nodular masses of olivine of a brilliant light green, and often as large as the fist. It is also very regularly columnar, and I observed it has not unfrequently happened that the seam separating two proximate columns cuts through one of the large imbedded knots of olivine, leaving a segment on either side. This fact seems to prove that the columnar divisionary structure was in its origin attended by a *powerful contractile force*, and also that it did not take place till the lava was so far consolidated, and the knots of olivine consequently so firmly compacted in its crystalline substance, as to separate along the line of the seam even when it divided them in two, sooner than quit their matrix.*

In its disposition, or, more properly speaking, its aspect, this bed differs from the one last described; for the river, instead of cutting a deep channel through the mass, and consequently exhibiting vertical sections of it on either side, generally flows over its surface, the upper and amorphous part of which it has worn away, and thus disclosed on many points a plane horizontal section, in which the polygonal extremities of the columns are united into a sort of pavement (called by the natives, as with us in the north of Ireland, *Pavés de Géans*, or Giants' Causeways), not

---

* In the face of such a fact as this, it is difficult to deny the reality of a *contraction*, or to speak of the columnar structure being occasioned by the "mutual *pressure* of spherical concretions."—See Scrope on Volcanos, p. 135 *et seq.*

unlike those of the Roman roads in Italy, but arranged with far greater neatness and accuracy of design.

The columns here, as throughout the Bas Vivarais, are usually hexahedral, often of five sides; those of four occur rarely, of seven still seldomer; I met with none of eight or nine; and of three, only when interposed between larger columns in the manner of a pyramidal wedge. The columns are habitually of small diameter, not often exceeding ten or at most twelve inches. They are sometimes divided by very frequent joints; at others attain a length of sixty feet without any separation. It struck me as a remarkable, and perhaps not a fortuitous coincidence, that, while the basaltic lava of Burzet is thickly sprinkled with knots of olivine, the granite, from the interior of which it has flowed in such abundance, contains an equal proportion of similarly shaped and sized nodules, composed of granular pinite with interspersed mica and quartz; this character prevailing only in a certain district near the site of the eruption. If the basaltic lava was derived from the fusion and recrystallization of granite, may we imagine the knots of pinite to have been converted during the process into olivine?

3. *Volcano of Thueyts.*—A volcanic cone to the east of the village of Thueyts is connected with that of Montpézat by a small portion of the primary ridge on which both eruptions broke forth, probably at the same epoch. It is much inferior in size to the other, and without a regular crater; but has vomited an abundant flood of lava into the bed of the Ardèche. Thueyts is built upon its surface, which is cellular and scoriaceous. The river has gnawed out a new channel between the bounding cliffs of this plateau and the granite of its southern bank, and exhibited a majestic colonnade of basalt, about one hundred and fifty feet in height, and extending with few breaks for a mile and a half along the valley.

4. The cone of Jaujac,* called *La Coupe de Jaujac*, from its cup-shaped crater, has this peculiarity, that it rises from a coal formation, occupying the bottom of a long transverse valley, between elevated ranges of granite and gneiss, and would thus appear to countenance the exploded notion that volcanic fires are alimented by immense beds of coal. The primitive fragments, however, frequently found enveloped by its scoriæ and basalt, sufficiently prove, if proof were wanting, that the source of the erupted matters existed, at least, below the sandstone which encloses the coal strata.

The crater is very large and regular, but breached towards the north. Its figure is elliptical; the longer axis being directed north and south. Its sides as well as those of the cone are thickly clothed with chesnut-woods; and it is remarkable here as elsewhere, that those trees which grow on the volcanic are much larger and more productive than those on the primitive soil around. The earth resulting from the trituration and decomposition of recent basalt seems to be peculiarly favourable to the vegetation of the Spanish chesnut. Those of the woody region of Ætna are a well-known instance of prodigious luxuriance. At the foot of the cone gushes a mineral spring strongly impregnated with carbonic acid gas.

From the northern breach of the crater may be traced a vast current of basalt, which occupies and descends the valley of the Alignon to a distance of between two and three miles. The village of Jaujac stands upon this bed, on the brink of a mural precipice, which is continued to the termination of the current, and everywhere presents a columnar range of almost unexampled beauty, about 150 feet in height.

The river has excavated its channel between this and the

---

* See Frontispiece..

granite of its western bank, but on the other side the bed of basalt, which is of considerable width, is closely soldered to the opposite range of primary rocks, whose débris, accumulating at their base into a cultivated slope, have concealed the line of junction. The lava of the Coupe de Jaujac either stopped without reaching the Ardèche, or, which is more probable, is confounded with that of another neighbouring cone, called—

5. *La Gravenne de Souillols*, which also disgorged its lava into the bed of the Alignon about 300 yards above the junction of this river with the Ardèche. A wide and massive plateau of basalt thus formed, after entering the valley of La Baume, prolongs itself to some distance below Niaigles, bordering the Ardèche on the south with a bold and precipitous wall which may be seen to rest on a layer of pebbles, the ancient bed of the river.

The face of this line of cliff, as well as that of the current of Jaujac, exhibits two distinct portions or stories separated by a well-defined straight line generally parallel to both surfaces; the upper consists of small, irregular, and matted columns; the lower, habitually occupying about one-third of the whole height, of large, perfect, and vertical pillars, which almost appear placed as an artificial support to the immense superincumbent entablature. On a close inspection the regular columns below are seen to proceed so immediately out of the upper and comparatively amorphous mass, that it is evident the whole was cast at once, and the two portions cannot be considered as different beds or currents.* Their singular difference of structure may be accounted for, I conceive, by supposing the upper part to have been first consolidated by exposure to the air, while the lower was still in motion, and flowing down the river channel beneath.

---

* See Frontispiece

When the efflux of lava from the vent finally ceased, this latter mass will have cooled down with extreme slowness, and been consolidated in a regular and tranquil manner, such as would facilitate the establishment of straight and vertical axes of contraction, and the production of very regular hexahedral columnar concretions, perpendicular, as usual, to the cooling surfaces.*
A view of part of this range near the Pont de Baume engraved in the work of Faujas, after due deductions for exaggeration in size and its faulty execution, gives not an inaccurate idea of the disposition of the columns.

It is here that the river, descending both from Burzet and Montpézat, joins the Ardèche; so that, if we suppose their eruptions to have been contemporaneous, or nearly so, it is not at all improbable that the five lava-currents we have described, flowing from as many different points, were once united on this spot, exactly as the rivers were, and still are, whose channels they usurped. In the bed of the river Ardèche, at and for some distance below this point, may be seen in summer, when the stream is inconsiderable, a number of articulations of basaltic columns, in which a nice observer may recognise the mineral characters of the different lava-currents of the tributary valleys. As you follow the course of the river, these columns show themselves less frequently and are more water-worn, till at the distance of a mile or two they are reduced to little more than rounded blocks, and nearly assimilated to the other boulders which cover the dry channel of the river. These basaltic boulders continue to diminish in size as you descend, and few are to be met with as far down the stream as Aubenas so large as a man's head; further on they are reduced to mere pebbles, and are no doubt still more comminuted before the Ardèche

---

* See Considerations on Volcanos, p. 136.

carries them with it into the Rhône. This observation illustrates the process by which both the basalt and granite that once filled these valleys have disappeared. A wintry flood undermines and detaches a prism of basalt from one of the columnar ranges. The next flood drives it on a few inches; or, if by its form and position it is enabled to roll without much difficulty onwards, a few feet. This operation is repeated year after year, and in the mean time, even when remaining stationary, it is exposed to the immense friction of all the smaller boulders and pebbles which are drifted over it by the ordinary as well as the extraordinary force of the current. By the continuance of this process it is at the same time carried forwards, reduced in size, and brought to approach to a globular form, the most favourable to its transport, in consequence of which the rapidity of its progress along the channel of the river is progressively accelerated, till, diminished to the size of gravel or silt, it is taken into complete suspension, and carried sooner or later in this state into the ocean.

At the foot of the cone of Souillols, near the hamlet of Neyrac, rises a spring strongly impregnated, like that of Jaujac, with carbonic acid gas; and from the bottom of a small hollow close by this gas emanates in such abundance that the exact phenomena of the Grotta del Cane are reproduced, and dogs or fowls may be stifled and revived *ad libitum*.

6. The last of the cones we have mentioned, called *La Coupe d'Ayzac*, rises on the ridge of one of the granitic abutments that project from the steep escarpment of the Haut Vivarais. It has a beautiful crater slightly broken down towards the north-west; and from the breach a stream of basalt may be seen to descend the flank of the hill, and turning to the north-east enter the valley of the river Volant, which has subsequently cut it entirely across, and disclosed three distinct storied ranges; the

lowermost very regularly columnar, that in the middle less so, and the upper nearly amorphous, cellular, and with a ragged scoriform surface.* This current, which appears originally to have occupied the whole bottom of the gorge in an extent of about four miles, from the village of Antraigues nearly to that of Vals, where the Volant flows into the Ardèche, has been worn away and carried off on many points by the violence of the torrent. Its relics adhere in vast masses to the granite rocks on both sides, particularly in the receding angles of the valley, sometimes bordering the river for a considerable distance, and reaching the height of 160 feet above it. The lower portion of this bed is very beautifully columnar, the upper obscurely so; this latter has been in parts destroyed, and a pavement or causeway left, formed by an assemblage of upright and almost geometrically regular columns fitted together with the utmost symmetry. The columns, as usual, are always at right angles to the containing surfaces, through which the heat escaped.

Such are the most interesting volcanic remains of the Bas Vivarais; they evidently form the continuation of the line of recent eruptions which we have noticed on the lofty table-land above.

The exceeding freshness of their scoriæ and the general regularity of their craters would seem to indicate their being of a later date than the others; but this state of superior preservation may to some extent be attributable to their minor elevation and more sheltered position; and the vast amount of excavation already accomplished by the erosive force of the mountain streams along the valleys which feed the Ardèche, since their invasion by currents of basalt, proves the origin of these, however recent in comparison with the beds of the same nature around the Mezen, to belong to an æra incalculably remote

---

* See the annexed drawing, Plate XV.

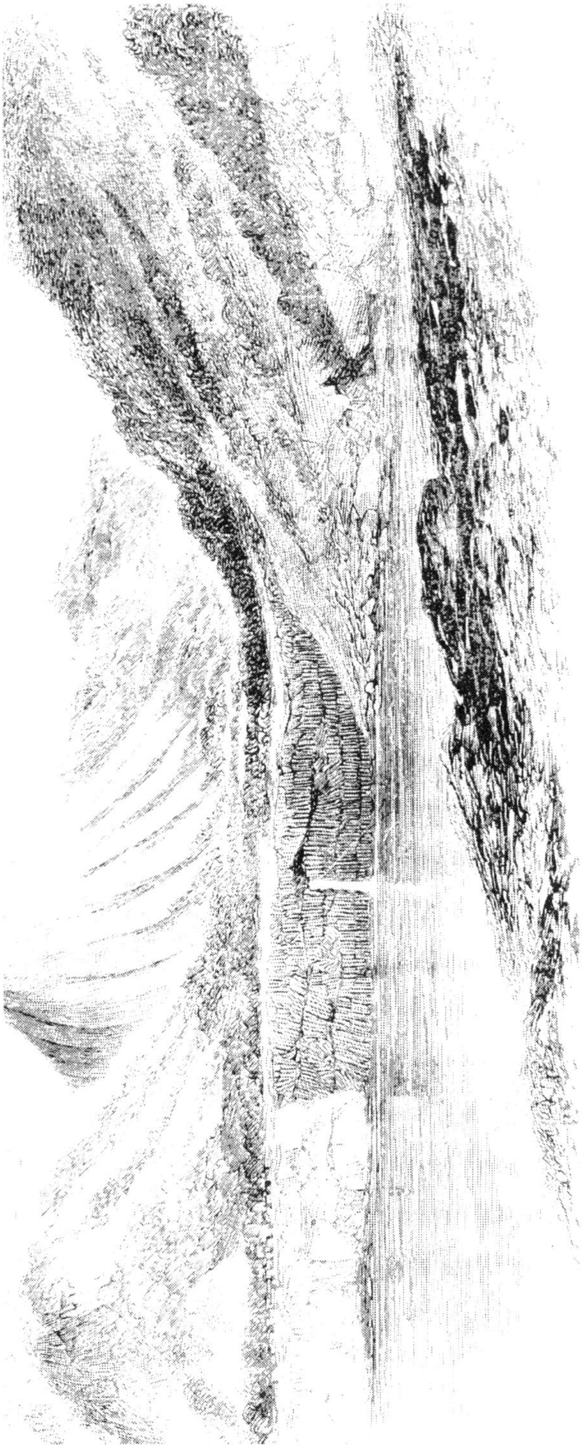

29 AP 58

with respect to ourselves. I am not aware of any organic remains having been observed under these lava-beds, which might throw any light on the geological age to which they belong.

There is one instance, in the immediate vicinity of Aubenas, in which a volcanic eruption appears to have taken place within, but at little distance from, the limits of the secondary limestone formation. The hill upon which this town is built consists of strata dipping to the south at an angle of 10°. Within a hundred yards north of the town walls, and scarcely so much from the sandstone on which they rest, these limestone strata are abruptly broken through by an enormous vertical dyke of basalt, which protrudes about 30 feet above the summit of the hill, and may be traced down its side into the valley of the Ardèche, pursuing an easterly direction. At the lower part of the hill the dyke measures from 12 to 20 feet between its cheeks; higher up it increases in bulk, and at the summit is 90 feet thick. From thence it makes a sudden curve to the north, returns to the north-east, and is again to be traced down the same face of the hill about a hundred yards from the other branch; thus including a large portion of the limestone strata, which preserve their parallelism and general inclination. This second branch of the dyke has dislocated and shattered the limestone within a space of from 40 to 50 yards, enveloping blocks and fragments of all sizes, with which it is mixed confusedly, and presents the appearance of a loose limestone breccia traversed or cemented by veins of basalt. Two smaller veins may also be observed cutting perpendicularly the undisturbed limestone strata enclosed by the two arms of the principal dyke. The one is 15, the other 10 inches thick.

The basalt itself is of a very dark grey colour, compact, fine-grained, hard, and tough; its fracture tends imperfectly to the conchoidal, and it breaks into curved and angular pieces.

It encloses numerous large crystals of greenish-black augite, and a brilliant sea-green olivine. It is almost universally penetrated by calcareous infiltrations, which line all the cavities of the cellular parts with snow-white calc-spar, and are even found in the interior of some of the crystals of augite. Small fragments also of limestone are enveloped by the basalt, to which they adhere firmly, the two substances appearing soldered together by an intimate though partial mixture. Occasionally the whole enclosed fragment of limestone is hardened, of a dark grey colour, a granular or crystalline texture, and a siliceous or cherty aspect; in some instances, on the contrary, it is white, earthy, and effervesces readily with acids; while in others the basaltic and calcareous particles seem to have united and separated again, the latter into small grains or patches resembling white porcelain imbedded in a dark grey basaltic base. The large masses of limestone in contact with the veins exhibit few or no signs of alteration.

The whole appearance of this dyke and its subordinate ramifications announces its having been forcibly propelled from below through the calcareous strata; but at what epoch, or whether connected in any way with the neighbouring eruptions whose products have been described above, it is difficult even to conjecture.

This is the most southern instance of any volcanic formation which attaches itself to the primary mountain-group of Central France. The interval from hence to the neighbourhood of Béziers, Agde, and Pezenas, or Aix and Toulon,—all points upon which volcanic remains are also to be found,—is, I believe, totally exempt from such substances; and it seems a very fair conclusion, that the enormous mass of secondary strata which from this spot southward covers the fundamental granite, and consequently the focus of all volcanic energy, stifled and impeded its ulterior development.

CHAPTER IX.

CONCLUDING REMARKS.

IF we now turn to take a wide and general view of the interesting tract whose particular formations have been described in the preceding pages, some of its most prominent features will be found to suggest considerations of no mean geological importance.

1. We first remark the peculiar position of the great mass of primary or Plutonic rocks, piercing, like a vast protuberance, through the secondary strata which surround it on every side, and appearing to have formed an island in the ocean from the commencement of the secondary period, and at no subsequent time to been covered by the sea; for,

2. It is observable that no marine deposits later than the Jurassic system are to be found within the area of this elevated district, or nearer to it than the low chalk hills of Champagne and Touraine on the north, and the still lower basin of Languedoc on the south; while in lieu of these a very massive calcareous, and in part arenaceous formation, the accumulated sediments of one or more extensive freshwater lakes, occupies the principal depressions in the primitive table-land, and prolongs itself from thence northwards as far as Nevers and Moulins.

3. Having noticed at least three of these calcareous freshwater deposits within the mountainous district to which our examination has been confined, the question naturally rises in the

mind, whence could such immense accumulations of carbonate of lime be derived? Other examples of analogous formations are usually found in cavities surrounded by chalk or secondary limestone. The calcareous formations of Auvergne, the Cantal, the Haute Loire, and Montbrison are entirely cased in granitic rocks. This circumstance wholly prevents our supposing the carbonate of lime to proceed from the detritus of other limestone strata; and it is therefore to the calciferous springs of St. Alyre, St. Nectaire, Rambon, Chalucet, and the numerous others of a similar nature which we can hardly doubt to have been more productive when the subterranean forces of this district were in greater activity, that we can alone look for its origin. A large proportion of the lime emitted from such of these sources as were below the level of the lakes was first probably secreted by the charæ and other aquatic plants growing from their bottom (of which numerous impressions are still found), and thence taken as food into the substance of innumerable mollusks, which by the gradual accumulation of their shells gave rise to the marly strata of these lake-basins; while on those spots where the springs produced the matter in abundance, or in the open air, beds of more or less compact semi-crystalline limestone, or travertin, were formed by its precipitation. This view is strongly confirmed by the admixture with the lime both of siliceous matter, which is known to be frequently deposited from thermal springs in a volcanic country, and also of gypsum: for it has been shown that eruptions were habitually taking place during the deposition of the calcareous strata of the Limagne and the Cantal, if not of the Haute Loire: and if sulphuretted hydrogen was evolved, as is usual during these phenomena, through the soft marly beds at the bottom of the lakes, the sulphuric acid uniting with the lime would necessarily produce that mineral: of which a part may have been precipitated on the spot, and the remainder carried away by the water into the lower lake-basins to be ultimately deposited there (Gypsum of Paris?). The strong odour of sulphuretted hydrogen emitted from the marly limestone of Le Puy seems indeed to prove that this gas was produced at the epoch of its deposition.

In fact, the calcareous freshwater formations of the centre of France differ but in this one respect (viz. the presence of gypsum and silex, which the thermal volcanic springs will account for) from the recent shell-marl deposits of the Bakie and other lochs in Scotland, described by Sir C. Lyell. In both are found Limneæ, Planorbes, Helices, a species of Cypris, the remains of Charæ and the Gyrogonites. In both, marly strata in which all traces of shells have disappeared, but which occasionally contain bones of mammalia and birds, alternate with others of a yellowish travertin limestone, often semi-crystalline, tubular, containing remains of vegetables, insects, &c., and with beds of sand. The basins of both still contain springs charged with carbonate of lime, and both occur in the neighbourhood of trap or volcanic rocks.*

4. With regard to the rocks of volcanic origin, they are distinguished primarily into, 1st. The products of three great habitual vents—the Monts Dore, Cantal, and Mezen, which appear to have been in activity towards the same time, and to have raged at intervals and with intense energy during a period of considerable duration. 2ndly. The products of single occasional eruptions from a vast number of separate apertures,

---

* I retain these remarks as they were printed in the first edition (1826), although aware that to the greater number of my readers such arguments will be familiar from the admirable development and confirmation they have since received in the works of my friend Sir Charles Lyell. See his Manual, p. 197-203, ed. 1855.

ranged closely along the line of what was probably a great compound fracture in the superficial granite, stretching from north-north-west to south-south-east, across the whole elevated tract, coincident with the general direction of its beds, and therefore with the axis of elevation. Both classes of volcanic products exhibit a great variety of mineral composition, from porous feldspathose trachyte to compact augitic basalt. Both rest indifferently upon the granitic rocks and the freshwater strata, and even in some instances are found to alternate with the latter, as for example, the trachyte of the Cantal near Aurillac, and the basalt of Gergovia, the Puy Dallet, Pont du Château, &c. The latter circumstance sufficiently proves eruptions to have taken place habitually from both the principal volcanos and the longitudinal fissure, during the period of the deposition of the freshwater formation. But it is equally certain that a large proportion of the basaltic lavas were erupted subsequently to the emptying of the freshwater lakes of the Limagne and the Haute Loire, since they have filled hollows and valleys eaten out of their stratified deposits.

An attempt has been made by many of the French geologists to ascribe the production of the volcanic rocks of Central France to two or three distinct periods alone, upon grounds either of peculiar mineralogical character, or their superposition to particular alluvial beds. It appears to me, however, that no classification in respect to age, founded on such data, can be relied upon. The most popular division perhaps is that of M. Rozet,* who refers them to three distinct epochs, which he calls " The Trachytic, Basaltic, and Lavic." But it is unquestionable, from the most direct evidence of superposition, that many of the basalts were produced as early as a large proportion of the tra-

---

* Bulletin XIV. p. 167.

chytes; and, on the other hand, many trachytes, for example the domitic puys, are so associated with the most recent cones of scoriæ and lava currents, that they must be considered of the same age: to which it may be added that the best defined trachytes are occasionally found to pass into clinkstone, and this again into basalt.* Some of the freshest lavas of the Monts Dôme, as those of Nugère and Volvic, are indeed almost wholly composed of felspar, and differ little if at all from many of the trachyte currents of the Mont Dore.† All, in truth, that can be said on this point is that, generally speaking, the eruption of trachyte preceded that of basalt, but that there were on several points alternate emissions of the two sorts of lava is not open to question; and this accords with what has been observed in other volcanic regions, where the general rule seems to have been the earlier production of trachyte, although exceptions are frequent.‡

As to the alleged uniform superposition of the basaltic currents to alluvia, containing rolled pebbles of basalt, as marking a particular age (which is the idea of M. Pissis §), it is quite certain that this feature is common to basalt of every age from nearly the first to the last produced; and necessarily so, since the lava stream will always have flowed into the channel of the then existing rivers, which must generally have contained alluvial gravel and rounded pebbles, the debris of the neighbouring slopes; in fact, such beds are found beneath what are evidently, from their circumstances of position, aspect, decomposition, and denudation, the oldest as well as the newest lava-currents. For these reasons I can admit no other criterion of the relative age

---

* See M. Raulin's paper on the Cantal. Bulletin XIV. p. 114.

† See note to p. 131, *supra*.

‡ In Iceland as well as Teneriffe trachytic domes and currents have burst through the earlier basaltic beds.

§ Bulletin XIV. p. 240, &c. 1843.

of the volcanic products I have described, than such as is derived from the circumstances above mentioned; especially, and before all, their position in relation to the levels, and the amount of degradation they, or the surrounding rocks, have evidently undergone since they were poured out as more or less liquid lavas upon the spots they now occupy. Upon this point some further remarks are necessary.

5. It is impossible to observe the many strips of the originally continuous freshwater formation which rise from the plain of the Limagne in long tabular hills, transverse to the general course of the river which now drains the principal valley-plain, without being convinced that each owes its preservation from the destruction which has swept away the remainder of the formation to its capping of basalt, which, by reason of its superior hardness, would naturally protect the underlying strata from the rains, frosts, and other meteoric agents, to which the uncovered intervals of the marly plain left by the emptying of the lake were permanently exposed. Such a capping, on the other hand, would afford a very inefficient protection against the denuding force of any violent deluge or general current of waters, to which some writers have attributed the excavation of the valleys intervening between these high basaltic platforms: more particularly as the direction of any such current passing over this district must have been coincident with the general direction of the valley of the Limagne, or from south to north; whereas the long strips and flat promontories in question *invariably* run east and west, preserving, as might be expected, the direction which was originally given by the lateral slopes of the valley-basin to the streams of lava that flowed into it from the heights on either side.

Again, had the whole excavation effected in the freshwater formation of the Limagne been produced *at once*, by any debacle accompanying an elevatory movement of the granitic base, or the eruption of some of the volcanos, or by any *diluvial* or other violent catastrophe, it is clear that the remnants of the lava-currents which had flowed into the freshwater basin *before* this epoch would be necessarily all found at one level, or nearly so, corresponding to the average level of the bottom of the lake-basin at that time; while, on the other hand, all the lava-streams which have flowed since the debacle or supposed deluge would be found at another nearly uniform, but much lower level, viz. that of the lowest places of the excavated valley. But, as we have seen, no marked distinction of this sort exists; no line can be drawn to separate the basaltic beds met with at high or low levels. They are found at all heights, from 1500 feet downwards, above the water-channels of the proximate valleys; and some even of the most distant in point of level are situated geographically close to one another.

Let us take, for instance, the two neighbouring basaltic platforms of Gergovia and La Serre, and the bed of basalt which occupies the bottom of the narrow valley that divides them, and which has flowed out of the recent vent marked by the Puy Noir.* Here are three long strips of basalt, each of which, by its gradual inclination in the direction of its greatest length, and by the positive remains of the cone of scoriæ in two instances at least out of the three, are proved to have flowed in a state of liquidity from volcanic vents opened upon the high granite platform, into the basin of the freshwater formation. Each must necessarily have occupied the lowest levels of that basin to which it had access; and therefore it is evident that, at the period when the lava of Gergovia flowed into its present position, there could have been no lower depression in the immediate vicinity. The

---

* See p. 91, and also Plate I., and the Map of the Monts Dôme.

hollow, therefore, into which the lava of La Serre flowed, must have been subsequently excavated, as that bed is everywhere —that is, on every line drawn transversely across both plateaux—at a lower level by from 300 to 500 feet than that of Gergovia, which runs parallel to it, and not above three miles distant. Again, by the same reasoning, it is evident that the intervening valley of Chanonat, the bottom of which is now covered by a still more recent bed of basalt, must have been excavated since the production of the lava of La Serre, which overhangs it at a height of more than 500 feet. Here then are three distinct steps in the process of excavation (it might be said *four*, for the rivulet of the valley of Chanonat has worn away a new channel, from twenty to fifty feet in some places below the substratum of the most recent of the three lava-currents), while according to the diluvian theory the whole process was simultaneous.

Moreover, in its mineralogical aspect the basalt of these different beds affords a confirmatory proof of their relative ages corresponding with their relative heights. That of Gergovia is compact, partly amygdaloidal, and much decomposed externally. That of La Serre has an appearance of much greater freshness, though still possessing the distinctive features of the older basalts. That of the inferior current of Chanonat is scarcely less recent in appearance than many currents of Ætna of which the date is known. Again, the cone and crater whence the last proceeded is entire and undisturbed. That which gave birth to the lava of La Serre has been much degraded, the heavier and more massive scoriæ, bombs, &c., alone remaining. The source of the basaltic current of Gergovia was probably the Puy de Berzé, which retains the character of a vent of eruption, but is still further worn down to a mere stump.

29 AP 58

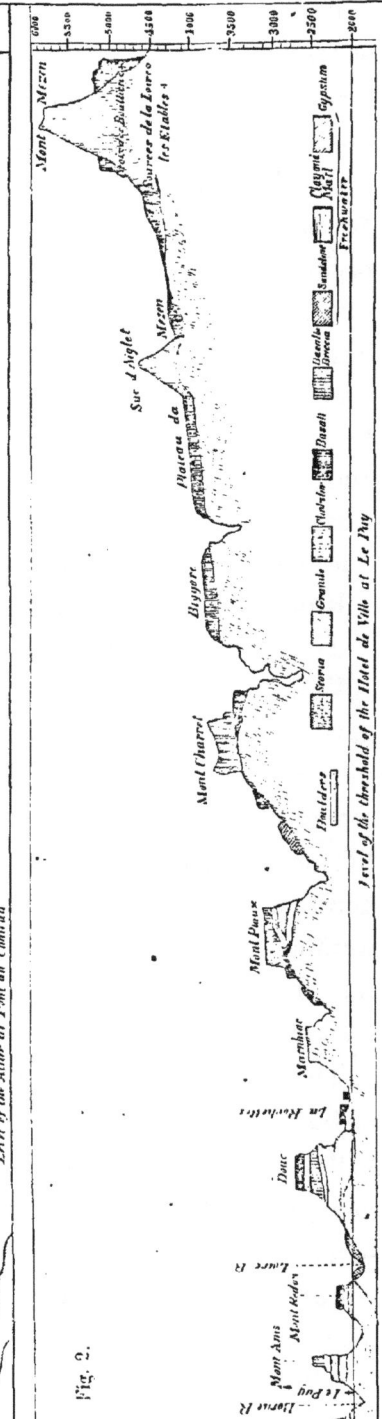

PROFILES OF THE PRINCIPAL HEIGHTS NEAR CLERMONT: PUY DE DÔME.

Fig. 1.

SECTION OF THE MONT MEZEN INTO THE FRESHWATER FORMATION OF LE PUY: N.W.–S.E.

Fig. 2.

Plate XVII.

Examples of this kind might be multiplied endlessly were it necessary, leading inevitably to the conclusion that the immense abstraction of matter which has occurred in the freshwater formation of the Limagne was for the most part effected *gradually* and *progressively*, and went hand in hand with the occasional flooding of parts of this valley and its tributary ravines by lavas emitted in the eruptive paroxysms of the volcanos on the neighbouring heights. Even were it allowable to have recourse to vague and hypothetical conjectures, we can conceive no *gradual and progressive excavating forces*, other than those which are still in operation wherever rains, frosts, floods, and atmospheric decomposition act upon the surface of the earth. To these agents, then, we must refer the effects in question, of which, with an unlimited allowance of *time*, no one will pronounce them to be incapable.

The composite sections of a part of the Limagne, seen in Plate XVII., in which the most remarkable currents of basalt that have descended at different epochs into this basin are given at their respective heights, will show how completely these volcanic remains mark in their actual position the progressive steps of this excavation, and exhibit a natural scale for measuring the duration of the process; a scale to complete which little is wanting but a knowledge of the intervals which elapsed between the successive eruptions by which these lava-currents were poured forth from the ridge of the Monts Dôme.

The lavas of the Bas Vivarais offer equally incontestable proofs of the same fact. We see there a number of deep and narrow valleys worn in the flanks of a steep range of granite, which have, at a certain epoch, been occupied, through a length of several miles, by lava poured in a liquid state from neighbouring volcanic vents, which has evidently filled them up to a high level, exactly as melted metal fills a mould into which it is poured. Since that epoch the valleys have been re-excavated in many parts to more than their former depth and width, the new channel being cut in some cases through the basaltic lava, in others through the granitic sides of the original valley. Now, if the first excavation of these valleys is to be accounted for by the hypothesis of a deluge, to what are we to attribute the second process? Not, most certainly, to a *second deluge;* for the undisturbed condition of the volcanic cones, consisting of loose scoriæ and ashes, which actually let the foot sink ankle-deep in them, forbids the possibility of supposing any great wave or debacle to have swept over the country since the production of these cones. The amount of excavation which has taken place subsequently to the epoch of these eruptions can then have been only effected by the streams which still flow there; and as this quantity bears a very considerable proportion to the extent of the original excavation, there can be no reasonable grounds for hesitating to attribute the latter to the same agency which effected the former; it being only necessary to assign a longer duration to the process to account for the difference in magnitude of the result. Those who, having before their eyes the proof that the immense quantity of solid rock removed from these valleys since their occupation by the lava has been effected by natural causes, such as are still in operation, will yet recur to a vague and unexampled hypothesis for the purpose of explaining the removal of a similar, or not very much greater, quantity, prior to that epoch, must do so in defiance of all the laws of analogical reasoning, by strict adherence to which we can alone hope to obtain the least acquaintance with those operations of Nature's laboratory of which we have not been actual eye-witnesses.

The volcanic district of the Haute Loire presents another chain of proofs equally conclusive of the same fact. It is im-

GRADUAL EXCAVATION OF VALLEYS.

possible to doubt that the present valleys of the Loire, and all its tributary streams within the basin of Le Puy,* have been hollowed out since the flowing of the lava-currents, whose corresponding sections now fringe the opposite margins of these channels with columnar ranges of basalt, and which constitute the intervening plains. Yet these lavas are undeniably of contemporary origin with the cones of loose scoriæ which rise here and there from their surface, and which would necessarily have been hurried away by any general and violent rush of waters over this tract of country. It is indeed obviously impossible that any such flood should have occurred; and we are therefore driven to conclude that the erosive force of the streams which still flow in these channels, together with the action of direct rains, frost, and other meteoric phenomena, have alone hollowed out this extensive system of deep, and, in some instances (as that of the Loire itself), wide valleys.†

---

* See Plate XI.

† [It is scarcely necessary to attempt a serious refutation of a species of quibble which has been too often brought forward in place of argument by the diluvian theorists; viz. that rivers are caused by the pre-existence of the basins through which they flow, and consequently these could not have owed their existence to the rivers that flow through them! It is clear that no extensive surface of the earth could at any time have been so uniformly smooth and level but that the rains falling upon it must have collected into streams as they drained off. The erosive force of these streams would necessarily by degrees excavate channels of a width and depth proportioned to the duration of the process, their magnitude and velocity, and the more or less destructible nature of the rocks over which they flow.

Of course, it cannot be doubted that vast irregularities of level, elevations and depressions of the earth's surface, on every scale of magnitude, have been occasioned by other causes, chiefly subterranean expansion. Mountains and valleys, in other words, the basins of our seas, lakes, and rivers, no doubt owe to circumstances of this nature their *primary* forms. Submarine currents, and the wearing action of the waves against litoral cliffs, will have scooped out many hollows in rocks exposed to these erosive forces; and convulsive oscillations of the ocean, or other aqueous reservoirs, occasioned by the sudden heaving up of large masses of the earth's crust, *may* have subsequently sent repeated waves over parts of our continents, the effect of which would be to open communications between distant basins, to create new and ex-

The time that must be allowed for the production of effects of this magnitude by causes evidently so slow in their operation is indeed immense; but surely it would be absurd to urge this as an argument against the adoption of an explanation so unavoidably forced upon us. The periods which to our narrow apprehension, and compared with our ephemeral existence, appear of incalculable duration, are in all probability but trifles in the calendar of Nature. It is Geology that, above all other sciences, makes us acquainted with this important though humiliating fact. Every step we take in its pursuit forces us to make almost unlimited drafts upon antiquity. The leading idea which is present in all our researches, and which accompanies every fresh

---

tensive denudations, and accumulate vast beds of transported fragments along the course of these mighty currents. But the proofs of the passage of such destructive deluges over any country require to be strictly made out and carefully tested. It will be seldom possible to show that the results cannot have been occasioned by the bursting of lake-basins, or other minor agencies still in operation, acting during an unlimited period. Before any just estimate can be formed of what share must be attributed to extraordinary catastrophes, and what to these minor but constant excavating forces, of the whole amount of change which has been evidently produced by the action of water in motion on the surface of the land, it is absolutely necessary to acquire a much more definite knowledge of the laws which regulate the circulation of water over the earth's surface, and its effect upon that surface, than we can at present be admitted to possess. It is too true that the greater number of geologists have sat down without hesitation to investigate by a sort of guesswork the origin of the changes and mode of production of the mineral masses which they observe on the surface of the globe, in complete ignorance, or at least with a total neglect, of those processes which are still daily employed by nature in the creation of fresh changes, and the production of new mineral masses on the same surface, bearing a complete analogy, to say the least of it, to the earlier phenomena, and older formations, which it is the business of geology to account for.]

This passage has been retained in the present edition, although the argument may appear unnecessary in the present advanced state of geological science, and especially after the publication of Sir C. Lyell's admirable 'Principles;' for the reason that neither my eminent friend, nor the bulk of geologists are, I think, even yet sufficiently impressed with the immense amount of excavation or denudation effected on supra-marine land by the erosive force of the pluvial and fluvial waters—in other words, by 'Rain and Rivers.' That story is yet to be written, but not by the eccentric author of the recent work under that title.

observation, the sound which to the ear of the student of Nature seems continually echoed from every part of her works, is—

Time!—Time!—Time! *

At least, since by a fortunate concurrence of igneous and aqueous phenomena we are enabled to prove the valleys which intersect the mountainous district of Central France to have been for the most part gradually excavated by the action of such natural causes as are still at work, it is surely incumbent on us to pause before we attribute similar excavations in other lofty tracts of country, in which, from the absence of recent volcanos, evidence of this nature is wanting, to the occurrence of unexampled and unattested catastrophes, of a purely hypothetical nature.

Nevertheless, although the evidence of their more or less degraded appearance and position proves the volcanic rocks of Central France to have been erupted at many different periods, and not at any one, two, or three epochs alone, yet there is reason to suppose some of these eruptive periods, more or less prolonged, to have been of extreme or paroxysmal energy. This, indeed, would only be in accordance with the habitual laws of volcanic action. To the earliest of such periods we may ascribe the production of the three chief volcanic mountains—the Mont Dore, Cantal, and Mezen. This outburst seems to have commenced towards the close of the deposition of the Miocene lacustrine strata. Several, likewise, of the independent basaltic beds

---

* It it very remarkable that, while the words *Eternal, Eternity, For ever*, are constantly in our mouths, and applied without hesitation, we yet experience considerable difficulty in contemplating any definite term which bears a very large proportion to the brief cycles of our petty chronicles. There are many minds that would not for an instant doubt the God of Nature to have existed *from all Eternity*, and would yet reject as preposterous the idea of going back a million of years in the History of *His Works*. Yet what is a million, or a million million, of solar revolutions to an Eternity?

that cap the highest hills of that formation, as well as those of calcareous peperino in the middle of the basin, belong evidently to a date quite as early. The larger proportion of the chain of puys of the Monts Dôme and the Haute Loire, especially the latter, must be attributed to another and comparatively recent eruptive era, which, from the evidence of organic remains found in the underlying alluvia, is referable to the Pliocene age. In the long interval between these two periods very many eruptions certainly took place from openings on the flanks of the volcanic mountains, as well as along the eastern zone of the Limagne and Forèz—eruptions whose lavas are found at varying heights above the existing river courses, and which therefore cannot be referred to any single epoch. And again, the most recent lavas and cones of the Mont Dôme and the Bas Vivarais have evidently broken forth at a yet later period, the Post-pliocene; probably even since the appearance of man in the country, although several species of the larger mammifers, now extinct, were contemporaneously its inhabitants.*

6. Another question of considerable interest presents itself, namely, what were the original limits of the several lake-basins of Central France, and the levels at which their waters stood?

In regard to those of the Upper Loire (of Montbrison and Le Puy) little difficulty exists, since the granitic barriers remain,

---

* A remarkable analogy exists on all these points between the several volcanic rocks of Central France and those of the district in Asia Minor, called by Strabo the 'Katakekaumene,' as was indeed noticed by Messrs. Hamilton and Strickland in their description of the latter. (Geol. Trans., vol. vi. 1816.) They likewise are shown to have been erupted at several distinct periods by the position occupied by the lava beds in relation to the adjoining valleys, excavated, as in Auvergne, through tertiary freshwater limestone; the earliest forming plateaux elevated 600 feet or more above the present river-beds, the latest occupying the channels of these streams (which, however, they have often cut through), while others occupy intermediate positions between these two extremes.

within which, if the narrow defiles that form the outlets of these basins were filled up, the superficial waters even now would rise to the highest levels at which any sedimentary strata are found. Such, however, is not the case of the Limagne and Cantal lake-basins. The tertiary beds of the first can be continuously traced from Brioude in the south to Decize near the confluence of the Loire and Allier in the north; and the lake would seem therefore to have been equally continuous. But in the vicinity of the latter spot we find no granitic or secondary heights capable of confining a body of water at a level of 2700 feet above the sea, which is the elevation attained by some of the lacustrine strata near Clermont and Issoire.* The highest hills now existing near the northern termination of this lacustrine formation do not attain 1000 feet.† It seems difficult to suppose a thickness of 1700 feet of rock to have been removed from the low hills north of Moulins, since the tertiary era; and it is more reasonable to imagine that changes of relative level have taken place, by which the southern portions of the tertiary formation may have been elevated, or the northern depressed.

M. Raulin, indeed, thinks he can trace the existence of a transversal east and west axis of elevation across the line of the Limagne, in the parallel of the Mont Dore and the Puy Barnère, the highest point at which the tertiary strata are left.‡ M. Pissis § is of opinion that the whole freshwater formation has been elevated since its deposition, so as to give it a slight general dip from west to east, and from south to north.

It appears unnecessary to decide exclusively for any one of these hypotheses, which may be all more or less accordant with fact. Although the absence of secondary strata through the

---

* At the Puy Girou, 2700 feet; at the Puy Barnère, 2730.
† Height of signal-post between Nevers and Magny (Oxford clay), 960 feet.
‡ Bulletin XIV. p. 587.
§ Bulletin, 2de Ser. p. 46, 1843.

entire granitic region proves it to have formed an island in the ancient ocean where those strata were deposited, it is quite possible that it may have sustained considerable absolute elevation since the formation of its tertiary beds, and especially during its eruptional era; indeed it would be unlikely that Central France should have remained unmoved during the elevation of portions of the neighbouring Alps and the basin of Switzerland from beneath the sea to a height of 5000 or 6000 feet. And again, with our knowledge of the repeated alternations of marine and freshwater deposits in the basin of Paris, into which the calcariferous waters of the Limagne overflowed, it is impossible to question the probability of occasional subsidence along the line of the intervening heights.

I am however of opinion that whatever changes of relative level took place, they operated over wide superficial areas, since few or no traces of disturbance are visible in the sedimentary beds of the Limagne. The surfaces over which the basaltic currents flowed, and which, having been since protected by them from the rain-fall, preserve the highest remaining surfaces of the lacustrine formation (although at that time they must have been its lowest levels), exhibit just that gradual inclination away from the bordering heights, where the greater number were erupted, which might be expected to prevail in the bottom of a shallow but gradually deepening lake. No sudden faults or dislocations appear to indicate any relative changes of level within areas of moderate extent.* On the other hand, the vast amount of

---

* This remark, however, must be confined to the Limagne. The freshwater strata of the Cantal have evidently suffered a certain amount of disturbance. M. Raulin describes the calcareous beds in the valley of the Alagnon as reaching a height of between 600 and 700 feet above the corresponding strata in the contiguous valley of the Cer; and some portions of these tertiary beds (viz. at Dienne) have been lifted up to an absolute elevation of 3680 feet—a position in which it is difficult to suppose they were deposited.

denudation to which the freshwater formation of the Limagne has been subjected, and which has left (as M. Ramond expressed it) only a few detached hills as relics of a former plain elevated many hundred feet above that now existing, may well be believed to have been accompanied by a corresponding destruction of superficial rocks at the northern end of the plain. I have attempted to show, not unsuccessfully, I hope, that a large proportion, if not the whole, of the degradation sustained since the tertiary period, as well by the granitic platform itself and its surrounding secondary zone, as by the freshwater strata, was effected by the slow and gradual but long-continued erosive force of the ordinary meteoric agents of denudation, rain, torrents, and river-floods, co-operating, as in this district they most probably did, at least during its eruptive periods, with frequent earthquake shocks, and perhaps a general elevation of the southern portion of the platform.

I cannot conclude the detailed description I have attempted to give of this interesting country, better than by quoting the eloquent summary of its characteristic features which Sir Charles Lyell has given.*

"We are here presented with the evidence of a series of events of astonishing magnitude and grandeur, by which the original form and features of the country have been greatly changed, yet never so far obliterated but that they may still, in part at least, be restored in imagination. Great lakes have disappeared—lofty mountains have been formed, by the reiterated emission of lava, preceded and followed by showers of sand and scoriæ—deep valleys have been subsequently furrowed out through masses of lacustrine and volcanic origin—at a still later date, new cones have been thrown up in these valleys—new lakes have been formed by the damming up of rivers—and more than one creation of quadrupeds, birds, and plants [Miocene, Pliocene, and Post-Pliocene] have followed in succession;

---

* Manual, p. 127, ed. 1855.

yet the region has preserved from first to last its geographical identity; and we can still recall to our thoughts its external condition and physical structure before these wonderful vicissitudes began, or while a part only of the whole had been completed. There was first a period when the spacious lakes, of which we still may trace the boundaries, lay at the foot of mountains of moderate elevation, unbroken by the bold peaks and precipices of Mont Dor, and unadorned by the picturesque outline of the Puy de Dome, or of the volcanic cones and craters now covering the granitic platform. During this earlier scene of repose deltas were slowly formed; beds of marl and sand, several hundred feet thick, deposited; siliceous and calcareous rocks precipitated from the waters of mineral springs; shells and insects imbedded, together with the remains of the crocodile and tortoise; the eggs and bones of water birds, and the skeletons of quadrupeds, some of them belonging to the same genera as those entombed in the Eocene gypsum of Paris. To this tranquil condition of the surface succeeded the era of volcanic eruptions, when the lakes were drained, and when the fertility of the mountainous district was probably enhanced by the igneous matter ejected from below, and poured down upon the more sterile granite. During these eruptions, which appear to have taken place after the disappearance of the [Lower Miocene] fauna, and partly in the [Pliocene] epoch, the mastodon, rhinoceros, elephant, tapir, hippopotamus, together with the ox, various kinds of deer, the bear, hyæna, and many beasts of prey ranged the forest, or pastured on the plain, and were occasionally overtaken by a fall of burning cinders, or buried in flows of mud, such as accompany volcanic eruptions. Lastly, these quadrupeds became extinct, and gave place to [Post-Pliocene] mammalia, and these, in their turn, to species now existing. There are no signs, during the whole time required for this series of events, of the sea having intervened, nor of any denudation which may not have been accomplished by currents in the different lakes, or by rivers and floods accompanying repeated earthquakes, during which the levels of the district have in some places been materially modified, and perhaps the whole upraised relatively to the surrounding parts of France."

APPENDIX.

CATALOGUE OF ORGANIC REMAINS, TABLE OF HEIGHTS,

AND

EXPLANATION OF THE
MAPS AND ENGRAVINGS.

APPENDIX.

ORGANIC REMAINS OF THE TERTIARY AND POST-TERTIARY FORMATIONS OF CENTRAL FRANCE.

VEGETABLE remains abound in the lower arenaceous beds of the lacustrine formations, as well as in the tuffs of the volcanic period, where they occasionally form beds of workable lignite. But in neither case as yet have they been scientifically determined. They chiefly consist of the leaves, fruits, and occasionally stems of dicotyledonous trees, or of reeds and other plants, the usual growth of marshy spots. The stems of charæ are very abundant as well as their seed-vessels. I am able to say little more of the fossil mollusks found in the same formations. The tertiary sandstones rarely contain shells, but some species of Cyrena have been found in them.* The associated or overlying limestones and marls abound in shells belonging to the genera Helix, Lymneus, Paludina, Bulimus, Cerithium, Cyrena, Unio, and Cypris. Some of the species mentioned by M. Bouillet† appear to be referable to an earlier period than that to which, on the best Palæontological authorities, we have considered the entire freshwater formation of Central France to belong, viz., the Lower Miocene. Sir Charles Lyell has recently touched upon this question in a supplement to his Manual.‡ And the matter remains still open to further investigation.

Two marine shells are said to have been discovered in a sandy stratum near Issoire, by MM. Bravard and Pomel, belonging to the genera Natica and Pleurotoma, and akin to some occurring in the Faluns of the lower basin of the Loire. Such a circumstance would seem to indicate either a reflux at some period of the waters of that river, or that the Miocene sea had actually ascended the

---

* Pomel, Bulletin, 2de Ser., vol. i. p. 579.

† Bull., vol. vi. p. 99 and 255.

‡ 1857, p. 10.

Allier. But it is not impossible to account for one or two such instances by supposing a few small mollusks living in the lower brackish waters to have been brought up the valley by birds preying on shell-fish. The bones of species belonging to the gull tribe are not unfrequent in the Auvergne fresh-water beds.

The Palæontologists who have most closely studied the Fauna of Central France are MM. Pomel and Aymard. As the lists given by these two authorities are not identical, and their opinions vary as to the divisions into which the several series should be classed, I think my readers will be more satisfied if, instead of myself attempting any comparative estimate of the two, I give both catalogues, with an abbreviation of the remarks by which their authors accompany them.

The following is M. Pomel's catalogue of the Fauna of the (Miocene) Lacustrine Strata of Central France :*—

I.—FAUNA OF THE LACUSTRINE STRATA OF CENTRAL FRANCE.

The species marked * have been found only in the basin of the Haute Loire; those marked † have been found both there and in the basin of the Allier likewise. Those without any distinguishing mark have been found in the latter (the Limagne) basin only.

MAMMALIA.

Order *Cheiroptera*.   *Locality.*

Species	Author	Locality
Palæonycteris Robustus	*Nob.*	Langy, near St. Gerard-le-Puy (Dept. Allier).

O. *Insectivora*.

Species	Author	Locality
Geotrypus Acutidens	*Nob.*	Cournon, Chaufours, near Issoire.
Antiquus	,,	Puy de Dôme (Coll. Laizer).
Galeospalax Mygaloides	,,	Marcouin, near Volvie.
Mygale Nayadum	*Pomel*	Chaufours, near Issoire.
Plesiosorex Talpoides	*Nob.*	Cournon and Chaufours.
Mysarachne Picteti	,,	Chaufours.
Sorex Antiquus	,,	Langy.
Ambiguus	,,	id.
Echinogale Laurillardi	,,	Mont Perrier, near Issoire.
Gracilis	,,	Antoing, near Issoire.
Erinaceus Arvernensis	*Blainville*	Cournon and Chaufours.
Nanus*	*Aymard*	Ronzon, near Le Puy.

---

* Pomel, Catalogue des Vertèbres Fossiles du Bassin Supérieur de la Loire, &c. Paris. 1854.

APPENDIX. CATALOGUE OF ORGANIC REMAINS.

O. *Rodentia.*   *Locality.*

Palæosciurus Feignouxi	*Nob.*	Langy.
Chalaniati	,,	id.
Steneofiber Escheri	,, (*Castor*)	id. and Chaufours.
Myoxus Murinus	,,	id.
Myarion Antiquum †		id., Cournon, Chaufours, Le Puy.
Musculoides	*Nob.*	Cournon.
Minutum †	,,	Chaufours (Le Puy ?).
Angustidens	,,	id.
Theridomys Breviceps	*Jourd.*	Perrier, Antoign, St. Yvoine.
Isoptychus Jourdani *	*Nob.*	Le Puy.
Aquatilis *	*Aym.*	id.
Vassoni	*Nob.*	Sauvetat.
Tæniodus Curvistriatus	,,	id.
Omegadus Echimyoides	,,	Chaufours.
Archæomys Arvernensis	*Laiz.*	Vaumas (Dept. Allier), Cournon, Chaufours, and Langy.
Palanæma Antiquus	*Nob.*	Cournon and Perignat.
Lagodus Picoides	,,	Langy.
Amphilagus Antiquus	,,	id. Volvic.

O. *Carnivora.*

Lutrictis Valetoni	*Pom.*	Langy, Gannet, Gergovia, Vaumas.
Plesiogale Angustifrons	,,	Langy.
Robusta	*Nob.*	id. Vaumas.
Waterhousii	,,	id. Cournon.
Mustelina	,,	id.
Plesictis Robustus	,,	id.
Gracilis	,,	id.
Croizeti	*Pom.*	id.
Lemanensis	*Nob.*	id.
Genetoides	*Pom.*	Cournon.
Palustris	*Nob.*	Langy.
Elegans	,,	id.
Amphictis Antiquus	,,	id.
Leptorynchus	,,	id.
Lemanensis	,,	id.
Herpestes Antiquus	,,	id.
Lemanensis	,,	id.
Primæva	*Pom.*	Vaumas.
Elocyon Martrides *	*Aym.*	Le Puy.

		Locality.
Cynodon Velaunum *	*Aym.*	Le Puy.
Palustre *	,,	id.
Canis Brevirostris	*Croizet*	Gergovia, Langy.
Amphicyon Brevirostris	*Nob.*	Langy.
Leptorynchus	,,	id.
Incertus	,,	id.
Crapidens	*Pom.*	id.

O. *Ungulata.*

Mastodon Tapiroides	*Cuvier*	Gannat.
Deinotherium Giganteum	*Kaup.*	Chaptuzat, Aurillac.
Cuvieri	,,	St. Germain Lambron.
Rhinoceros Lemanensis	*Nob.*	Billy, Vichy, Gannat, Chaptuzat, Le Puy, Bournon de St. Pierre.
Croizeti	,,	Vaumas, Gannat, Bansac.
Paradoxus	,,	Gannat, Vaumas, Perrier.
Palæotherium Magnum *	*Cuv.*	Le Puy (Aymard).
Gracile *	*Aym.*	id. id.
Velaunum †	*Cuv.*	Le Puy, Ronzon, Bournon de St. Pierre, near Brioude.
Duvalii *	*Nob.*	Ronzon.
Plagiolophus Ovinus *	,,	Le Puy.
Minor *	*Pom.*	id.
Tapirus Porrieri	,,	Vaumas.
Palæocherus Major	,,	Langy.
Waterhousii	*Nob.*	Perignat.
Typus	*Pom.*	id. Langy.
Suillus	*Nob.*	Langy.
Elotherium Aymardi	,,	Ronzon.
Ronzoni	,,	id.
Antracotherium Magnum	*Cuv.*	Issoire, Cournon, Chaufour, Vaumas, Digoin.
Cuvieri	*Pom.*	St. Germain Lambron.
Ancodus Velaunus †	*Nob.*	Ronzon, Vaumas.
Leptorynchus †	,,	id. id.
Incertus †	,,	Le Puy.
Aymardi *	,,	Ronzon.
Synaphodus Gergovianus	,,	Gergovia.
Cænotherium Laticurvatum	*Pom.*	Langy.
Metopias	*Nob.*	id.
Commune †	*Bravard*	Cournon, Le Puy, Chaptuzat.
Elegans	*Pom.*	Langy.
Leptognatum	*Nob.*	Chaptuzat, Cournon.

                                              Locality.

Cænotherium Geoffroyi .. ..   *Nob.*   .. ..   Langy
            Gracilis .. ..   *Pom.*   .. ..   Vaumas.
Lophiomerix Chalaniaci.. ..   *Nob.*   .. ..   Sauvetat, Cournon, Apt.
Dremotherium Traguloides ..   ,, ...   Langy.
            Feignouxi ..   *E. Geoff.* ..   id., Cournon, Chaufours.
Amphitragalas Elegans .. ..   *Pom.*   .. ..   id.
            Lemanensis ..   *Nob.*   .. ..   id.
            Communis* ..   *Aym.*   .. ..   Ronzon.
            Boulangeri ..   *Nob.*   .. ..   Lanzy.
            Meminoides..   ,, ...   id.
            Gracilis .. ..   ,, ...   id.

O. *Marsupialia.*

Hyænodon Leptorynchus † ..   *Laizer*   .. ..   Cournon, Sauvetat, Le Puy.
            Laurillardi ..   *Nob.*   .. ..   Antoing, near Issoire.
Didelphis Arvernensis † .. .. *Gerv.*   .. ..   Langy, Cournon, Sauvetat,
                                                    Le Puy.
            Crassa* .. .. ..   *Aym.*   .. ..   Le Puy.
            Antiqua .. .. ..   *Nob.*   .. ..   Cournon.
            Lemanensis .. ..   ,, ...   Sauvetat.
            Minuta* .. ..   *Aym.*   .. ..   Le Puy.

AVES.

M. Pomel leaves the numerous remains of birds found in this district undetermined, mentioning only the genera Phænicopterus, Anas, Ardea, and one resembling Numenius, several of the orders Rapaces, and Gallinacei, one of the last being as large as a peacock.

REPTILIA.

O. *Chelonia.*

Testudo Hypsonota .. .. ..   *Nob.*   .. ..   Langy, Bournouele.
            Lemanensis.. .. .. *Brav.*   .. ..   id.   Cournon.
Ptychogaster Heckei .. ..   *Nob.*   .. ..   id.   Chaptuzat.
            Emydoides ..   *Pom.*   .. ..   id.
            Abbreviata ..   *Nob.*   .. ..   id.
Chelydra Meilheuratiæ .. ..   ,, ...   Vaumas, Chaufours.
Trionyx .. ..   *Geoff.*   .. ..   id.   Chignat.

O. *Sauria.*			*Locality.*
Diplocynodus Ratelii †	*Pom.*		Ronzon, Langy, Chaptuzat, Perrier, Antoing, Sauvetat.
Varanus Lemanensis	*Nob.*		Chaufours.
Dracænosaurus Croizeti	,,		Cournon.
Sauromorus Ambiguus	,,		Langy, Marcouin, near Volvic.
Lacertinus	,,		id.
Lacerta Antiqua	,,		Cournon.

O. *Ophidia.*			
Ophidion Antiquus	*Nob.*		Langy.

O. *Batrachia.*			
Batrachus Lemanensis	*Nob.*		Langy, Cournon, Chaufours.
Nayadum	,,		Chaufours.
Lacustris	,,		id.
Protophrynus Arethusa	,,		id.
Chelotriton Paradoxus	,,		id. Langy.

PISCES.

O. *Ctenoides.*			
Perca Lepidota	*Agassiz*		Vichy, Gergovia.

O. *Cycloides.*			
Cobitopsis Exilis	*Nob.*		Chadrat, near St. Amand.
Lebias Cephalotis	*Agass.*		Corent.
Perpusilus	,,		Laps.

It appears from this list that the Fauna of the Miocene lacustrine strata, known to M. Pomel, comprehends the following number of species:—Cheiroptera, 1; Insectivora, 12; Rodentia, 18; Carnivora, 27; Ungulata, 42; Marsupialia, 7; Chelonians, 7; Saurians, 6; Ophidians, 1; Batracians, 5; Fish, 4: in all 130 species, besides several of birds as yet undetermined. Twelve species out of this number are found in the basin of Le Puy only; and eight are common to the basins of the Allier and the Loire (*i. e.* the Limagne and of Le Puy). Some of the most characteristic species, as, for example, all those which belong to the locality of Vaumas (Dept. Allier, not far from the confluence of the two rivers, Allier and

Loire), comprehending *Ancodus, Antracotherium, Tapir, Rhinoceros,* and *Chelydra,* associated with *Archemys, Amphictis, Herpestes, Amphicyon,* Cænotherium, Testudo, and Crocodilus, occur in the sands and sandstones which form the lowest beds of the lacustrine series. On the other hand, the locality of Langy, near St. Gerard-le-Puy (Dept. Allier), so rich in Cheiroptera, Insectivora, Rodentia, Carnivora, Ungulata, as also in birds, Saurians, Tortoises, Lacertæ, Ophidia, and Batracia, belongs to the Indusial limestone, which generally appears in the upper beds of the series. Chaptuzat, Marlouis, and Gannat in the Puy de Dôme are in the same position.

From these and other observations, M. Pomel concludes that the entire series of lacustrine beds of the two basins belong to the same Geological and Zoological period, and cannot be distinguished chronologically by their Palæontological characters; the same species being found in the most recent as in the most ancient beds of the series; and whatever differences exist being capable of reference to accidents of distribution or of discovery.

Comparing this Fauna with those of other European tertiary districts, M. Pomel finds it to bear the strongest analogy to that of Mayence, and he classes it therefore as earlier than the Falunian (Upper Miocene), and more recent than the Parisian Gypseous deposits (Eocene). This would establish it as of the "Lower Miocene" period, which in fact is the position, as already stated, assigned to it by Sir C. Lyell.

II.—PLIOCENE FAUNA OF THE VOLCANIC ÆRA OF CENTRAL FRANCE.

(Tuffs and Breccias of Mont Perrier, Cussac and Violette, (Haute Loire,) &c.

MAMMALIA.

O. Rodentia.		Locality.
Castor Issiodorensis	*Croizet*	Pumiceous Tuff of Perrier.
Arvicola Robustus	*Nob.*	id.
(?)		id.
Hystrix (?)	*Croizet*	id.
Lepus Lacostii	*Nob.*	id.

O. Carnivora.   *Locality.*

Ursus Arvernensis	*Croiz.* and *Jobert*	Pumiceous Tuff of Perrier.
Lutra Bravardii	*Pom.*	id.
Mustelina	*Nob.*	id.
Zorilla Antiqua	(*Rabdogale Nob.*)	id.
Felis Arvernensis	*Croiz.* and *Job.*	id.
Pardinensis	,,	id.
Brachyryncha	,,	id.
Issiodorensis	,,	id.
Brevirostris	,,	id.
Incerta	*Pom.*	id.
Meganthereon Cultridens	*Nob.*	Under Basalt at Sainzelles, near Polignac.
Macroscelis	*Brav.*	Tuff of Perrier.
Hyæna Perrierii	*Croiz.* and *Job.*	id.
Arvernensis	,,	id.
Dubia	,,	id.
Vialetti*	*Aymard*	Vialette (Haute Loire).
Canis Megamastoides	*Pom.*	Tuff of Perrier.

O. Ungulata.

Mastodon Arvernensis †	*Croiz.* and *Job.*	Tuff of Perrier, Vialette, Mirabelle (Ardèche).
Borsoni † (M. Vellavus, *Aym.*)	*Kays*	Tuff of Perrier, Vialette, Le Puy.
Rhinoceros Elatus	*Croiz.* and *Job.*	Mont Perrier.
Tapirus Arvernensis	,,	id.
Sus Arvernensis	,,	id.
Cervus Roberti*	*Job.*	Polignac, near Le Puy.
Perrierii	*Croiz.* and *Job.*	Tuff of Mont Perrier.
Issiodorensis	,,	id.
Etueriarum	,,	id.
Pardinensis	,,	id.
Rusoides	*Nob.*	id.
Ardens	*Croiz.* and *Job.*	id.
Cladocerus	*Nob.*	id.
Ramosus	*Croiz.* and *Job.*	id.
Solilhacus*	*F. Robert*	Solilhac, near Le Puy.
Cusanus	*Croiz.* and *Job.*	Mont Perrier.
Leptoceros	*Nob.*	id.
Platyceros	,,	id.
Furcifer	,,	id.
Antilope Antiqua	,,	id.
Bos Elatus	*Croiz.*	id.
Elaphus	*Nob.*	id.

The Fauna of the Pliocene period thus contains of Rodentia 5, Carnivora 17, Ungulata 23 species: in all, 45. Neither Chelonians nor Saurians have been hitherto discovered in it. The lakes had been evidently drained.

It will be observed that this Pliocene Fauna is almost wholly confined to the tufaceous conglomerates of one locality, that of Mont Perrier near Issoire.* Sir Charles Lyell's paper on the subject of this remarkable deposit, read before the Geological Society, November 19, 1845, will be well known to my readers. The freshwater strata and the overlying sheets of basaltic lava had evidently been eaten into by deep valleys before the deposition of these tuffs, and the gravel beds on which they rest. Sir C. Lyell distinguishes two distinct beds containing bones interstratified with the tuffs. Subsequently to this eluvial accumulation of trass-like tuffs from the Mont Dore new valleys appear to have been excavated through them, in which newer ossiferous alluvial deposits occur. And it is to these alluvial beds that a large part of the following (Post-Pliocene) Fauna from M. Pomel's catalogue belongs.

M. Pomel observes that the most striking characters of the Pliocene Fauna given above are the large assemblage of *Cervi*, the number of large Felis, of which one, the Megantheeron, is, besides the Mastodon, the only extinct genus of this Fauna. The Rodentia are of European genera. The Ungulata comprehend *Mastodon*, rhinoceros, tapir, sus, cervus, antelope, and bos. The elephant, hippopotamus, and horse are absent. M. Pomel considers this deposit as of the age of the Subapennine tertiary strata.

FISH OF THE BASIN OF MENAT.—*Qu.* POST-PLIOCENE?

Perca Augusta .. .. .. *Agassiz.*
Cyclurus Valenciensii .. .. ,,
Pœcilops Breviceps .. .. *Nob.*
Esox . . . ?

---

* See page 135.

III.—Fauna of Ancient Alluvia (Post-Pliocene).

This series (as will be seen below) combines two or more distinct periods.

Sir C. Lyell, indeed, distinguishes these alluvia in the neighbourhood of Issoire into four successive divisions (Nos. 6, 7, 8, 9, of his section.—*Manual*, ed. 1855, p. 552.)

MAMMALIA.

O. Insectivora. — *Locality.*

Talpa Fossilis	*Nob.*	Bone Breccia of Condes near Issoire and Neschers.
Sorex Exilis	,,	id.
Fossilis	,,	id.
Myosictis Fodiens	*Pom.*	id.
Musaraneus Priscus	*Nob.*	id.
Erinaceus Major	*Pom.*	Alluvium at Peyrolles near Issoire.

O. Rodentia.

Sciurus Ambiguus	*Nob.*	In crevices of Lava of Graveneire.
Spermophilus Superciliosus	*Kaup.*	Paix near Issoire, Coudes, and Neschers.
Arctomys Lecoq	*Nob.*	Champeix, Châtelperioux, and crevices of Lava of Graveneire.
Castor Faber?	*Lin.*	Near Clermont.
Myoxus Nitella?	,,	Coudes.
Arvicola Antiquus	*Nob.*	id. Neschers, Langy.
Pseudoglareolus	*Pom.*	Coudes.
Arvaloides	*Nob.*	id. Neschers, &c.
Joberti	,,	Coudes.
Lemmus Fossilis.		
Mus Sylvaticus	*Lin.*	id.
Cricetus Musculus	*Nob.*	id.
Lagomys Spelæus?	*Owen*	id.
Lepus Diluvianus?	*Pietet*	id. Aubière, Champeix, Neschers, Châtelperron.
Cuniculi Affinis		id.

O. Carnivora.

Ursus Spelæus	*Blum.*	Caves of Châtelperron, Champeix, and Montaigu le Bellin.

APPENDIX.   CATALOGUE OF ORGANIC REMAINS.   227

		Locality.
Meles Fossilis	*Auct.*	Caves of Châtelperron, Champeix, and Montaigu le Bellin.
Mustela Schmarlingii	*Nob.*	Coudes, Aubières, Neschers.
Putorius Fossilis	*Auct.*	id.
Gale	*Nob.*	id.
Microgale	,,	id. Neschers.
Macrossoma	,,	id.
Felis Lyncoides	,,	id. Tour le Boulade.
Minuta	,,	id. Aubières.
Spelæa	*Goldf.*	Montaigu Cavern.
Meganthereon Latidens *	*Nob.*	Sainzelles, Haute Loire.
Canis Spelæus	*Goldf.*	Coudes, Tour de Boulade, Montaigu.
Neschersensis	*Croiz.*	Neschers.
Vulpes Fossilis	*Auct.*	id. Coudes, Aubière, Châtelperron, Sainzelles?
Hyæna Spelæa †	*Goldf.*	Scoriæ of St. Privat d'Allier (Haute Loire).
Brevirostris †	*Aym.*	Sainzelles, Ardes near Issoire.

O. *Ungulata.*

Elephas Meridionalis	*Blum.*	Tour de Boulade, Vichy, &c.
Primigenius	*Nesti*	Malbattu, Clermont, Le Puy, &c.
Priscus	*Goldf.*	Plain of Sarliève.
Rhinoceros Leptorhinus	*Cuv.*	Peyrolles near Issoire, Malbattu.
Aymardi *	*Nob.*	Under Basalt (Haute Loire).
Theiorinus	,,	Tour de Boulade, Vichy, Châtelperron.
Equus Adamiticus	*Schl.*	Paix, Boulade, Coudes, Neschers, Gergovia, Le Puy, &c.
Robustus	*Nob.*	Champeix, Malbattu, Peyrolles, upper beds of Perrier.
Tapirus Elegans †	,,	Le Puy, Tonneil.
Sus Priscus	*M. Serres*	Coudes, Boulade, Montaigu, Châtelperron caverns, &c.
Hippopotamus Major †	*Cuv.*	St. Yvoine, Tormeil, Montaigu, Sainzelle near Le Puy.
Cervus Guettardi †	,,	Boulade, Nerchers, Coudes, St. Yvoine, Châtelperron.
Somonensis	,,	Gergovia.
Intermedius †	*M. Serres*	Boulade, Champeix, Châtelperron, Scoriæ of St. Privat, at Le Gard.
Macroglochis	*Nob.*	Peyrolles near Issoire.
Ambiguus	,,	id.

			Locality.
Antilope Aymardi *		Nob.	Boulade, Le Puy.
Incerta		,,	Coudes.
Ovis Primæva		Gerv.	Coudes, Châtelperron.
Capra Rozeti		Pom.	Malbattu.
Bos Primigenius †		Blum.	Boulade, Champeix, Aubières, Vichy, Châtelperron.
Giganteus *		(Velaunus?)	Cussac.
Priscus †		Schlot.	Peyrolles, Tormeil, Anciat.

O. *Reptilia.*

Lacerta Fossilis		Nob.	Neschers.
Coluber Gervasii		,,	Coudes.
Fossilis		,,	id.
Rana Fossilis		,,	id.

It has been already mentioned that M. Pomel refers this Alluvial Fauna to two distinct periods, the earliest of which is characterised by exclusively containing the following species:—*Erinaceus Major, Ursus Spelæus (Neschersensis, Croizet), Hyæna Brevirostris, Canis Neschersensis, Elephas Meridionalis, Rhinoceros Leptorhinus, R. Aymardi, Megantlereon latidens, Tapirus, ... ? Equus Robustus, Hippopotamus Major, Cervus Ambiguus, C. Macroglochis, Capra Rozeti, Bos Priscus.* The other species in the catalogue are for the most part common to these earlier, and to several later alluvial deposits, occurring either in the plain of the Allier at no great height above its present level, or in taluses leaning against the sides of existing valleys, or in gravel beds supporting the more recent lava currents, or in caves and fissures of these or the older rocks. M. Pomel does not recognise any distinction of age founded on Palæontological evidence between these different last-mentioned deposits.

The greater number of *genera* represented in them are still found in France, and perhaps some of the *species*. Those not now existing in the country are *spermophilus, arctomys, lemur, cricetus, lagomys, ursus, hyæna, elephas, rhinoceros, antilope*. But even of these many species are still to be found in Europe. He remarks as singular that together with the elephant, rhinoceros, and hyæna, which recall a warmer climate than that of France in the present day, occur marmots, lemmings, bears, and others now for the most part inhabiting Alpine or colder climates. And he observes in explanation of this, that the molars of the herbivorous animals of this Fauna show marks of their having largely fed upon coniferous plants, leading to the opinion that this district in their time possessed a climate colder than the actual one.

This Fauna is in fact almost identical with that of the older superficial alluvia or drift of the entire surface of Europe.

I have already shown that, in company with this last-mentioned group of species now extinct, human remains have been discovered in the volcanic tuff of Le Puy.

M. Aymard, Vice-President of the Academic Society of Le Puy, and a naturalist who has attentively studied the Geology and Palæontology of the Haute Loire, and possesses a very rich collection of its fossils, classes the tertiary formation of that district in the following manner: *—

Lacustrine Strata.
1. *Lower Eocene.*—Variegated clays and marls of Le Puy, Emblavès, and Brioude.
2. *Upper Eocene.*—Gypseous marls and clays of the basin of Le Puy.
3. *Lower Miocene.*—Calcareous marls—siliceous limestone, and sand beds of the basins of Le Puy and Brioude, St. Pierre Eynac, Mathias near Fay-le-froid, &c. &c.

4. *Pliocene*—comprehending three successive series of Alluvial deposits, all underlying the volcanic rocks, viz. A. the inferior sand beds of Vialette, Picheviel, Taulhac, Laroche near Vals, Coupet, &c. B. Intermediate alluvial beds and stratified breccias of Sainzelle, near Polignac, &c. C. Upper alluvia and breccias, tuffs, &c., of Solilhac, Polignac, Le Collet, Estronilhas, St. Privat d'Allier, Chilhac, Montredon, Cussac, Les Trois Pierres, Corsac near Brivés, &c.

5. Post-pliocene or superficial alluvia and volcanic stratified breccias, ossiferous fissures, detritus, &c., of Denise, Croix-de-la-Paille, Malpas near Aiguille, &c.

The organic remains contained in these divisions, according to M. Aymard, are as follows:—

No. 1.—A few herbaceous plants, no trace of animals, except perhaps some bones of the *Palæotherium Primævum.*

---

* In a paper read by him before the Scientific Congress of France at Le Puy, 13th Sept. 1855. See the Report, vol. i. p. 228.

No. 2.—Two species of mammalia, viz., *Palæotherium Subgracile*, Aym., and *Monacrum Velaunum*, Aym., the eggs of some large aquatic bird, impressions of small fish, one crustacean *Elesilphus limosus*, and a few freshwater mollusks, lymnei, paludinæ, planorbes, and cyclades.

No. 3 is much richer in fossils, a single bed at Ronzon near Le Puy containing the greater part of the following list of new species all named by himself. *Insectivora*:—Tetracus Nanus. *Carnivora*:—Cynodon Velaunus, C. Palustris, Elocion Martrides. *Rodentia*:—Theridomys Aquatilis and T. Jourdani, Myoterium Minutum, M. Aniciense, Decticus Antiquus, Elomys Priscus. *Pachyderms*:—Ronzotherium Velaunum, R. Cuvierii, Palæotherium Gervaisii, Paloplotherium Ovinum, Entelodon Magnus, E. Ronzoni, Bothriodon Platorynchus, B. Leptorynchus, B. Velaunus, Zooligus Picteti, Gelocus Communis, G. Minor, Palæon Riparium, Lathonus Vallensis. *Ruminants*:—Orotherium Ligeris. *Subdidelphis*:—Hyænodon Leptorynchus of Laizer. *Didelphis*:—Peratherium Elegans, P. Crassum, P. Minutum. Of Birds he recognises the remains of 15 species not well determined as yet, chiefly aquatic, such as cranes, flamingoes, plovers, gulls, with some of the order of *Rapaces*. Of Reptiles several Chelonia, Chersites, Elodites, and perhaps Potamites; several large and small Saurians, and more than one Batracian. Of Fish one only species, small but very numerous, Pachystelus Gregatus, Aym. Of Insects, 2 species of Coleoptera, and several belonging to marsh-frequenting genera. Two species of Crustacea, Elosilphus Limosus and Cypris Faba; several Palustrine mollusks, of the genera Lymneus, Planorbis, Helix, Cyclas, &c.; Infusoria, and numerous impressions of the leaves of dicotyledonous and leguminous plants, Comptonia, and Chara (Gyrogonites).

There appears on examination of these catalogues scarcely sufficient Palæontological ground for the triple division of the lacustrine formation which M. Aymard suggests; and it would seem more reasonable to refer the whole to a single period—the Lower Miocene, as is done by M. Pomel and Sir Charles Lyell.

Again, M. Aymard's division of his 4th class into three series seems an equally unnecessary complication. It corresponds to the Pliocene or older Alluvia of Pomel and Lyell. M. Aymard's observations on the character of its organic remains are valuable from his intimate acquaintance with the localities. They are all what he calls sub-volcanic, that is, contemporaneous with the volcanic products of the district, generally interstratified with the basaltic plateaux, or their underlying tuffs and breccias. A complete break separates the Palæontology of these Pliocene alluvia from that of the Lacustrine strata already described. They contain scarcely any of the genera, and *not one of the species*, hitherto found in any part of the last-mentioned formation. And it is not unreasonable to refer this great change to the occurrence in the interval of the earlier volcanic convulsions, and the consequent drainage and partial excavation of the lacustrine beds. New races of beings had succeeded to those in existence during the Lacustrine period, and many of them continued to inhabit the district even after the extinction of the most recent volcanos. The Flora of the volcanic period is represented chiefly by impressions of leaves and fruit belonging to trees still growing in the neighbourhood. The same may be said of its Fauna, so far as respects Mollusks, Insects, and Batracians. The Mammalia alone exhibit a very decided difference from those of the present age. It contains at least three species of mastodons,* a machairodus, a tapir, hyænas, a rhinoceros, elephants, hippopotamus, besides several extinct species of Canis, Cervus, Antilope, Bos, Equus, and Sus.

M. Aymard's 5th class, the Post-pliocene superficial alluvia, ossiferous fissures, &c., agrees with the parallel division of M. Pomel. Its Fauna is characterised by the first appearance of the Bear, and some small extinct species of Cervus, and also comprises some extinct species of elephant, rhinoceros, horses, and cervus, identical with those occurring in the earlier series. Man likewise (as we have seen) appears to have been an inhabitant of this district before the complete termination of this last volcanic period.

---

* One found at Vialette, Mastodon Vellavus, Aym., is larger by one-third than the M. Giganteus of Cuvier.

TABLE OF HEIGHTS

IN THE

VOLCANIC DISTRICT OF CENTRAL FRANCE.

[N.B. The greater number of these heights are extracted from Ramond's *Nivellement Barométrique des Monts Dore et Dôme*; the remainder are derived from articles in the *Journal des Mines*, by M. Cordier, &c., and from the work of M. Bertrand Roux.]

			Eng. feet above the Sea.
Puy de Sancy, Monts Dore	Trachyte	*Ramond*	6217
Plomb du Cantal	Clinkstone	,,	6096
Le Mezen, Haute Loire	Clinkstone	*Cordier*	5820
Col de Cabre, Cantal	Clinkstone	,,	5545
Le Mont Lozère	Mica-slate	,,	5535
Le Puy Mari, Cantal	Clinkstone	,,	5444
Pierre-sur-Haute, Forèz	Granite	*Ramond*	5409
Le Puy Violan, Cantal	Clinkstone	,,	5228
Puy Gros, n. of Baths, Monts Dore	Trachyte	,,	5040
Puy de Dôme	Trachyte	,,	4842
Croix Morand Plateau	Basalt	,,	4703
Source of the Loire, Haute Loire	Clinkstone	*B. Roux*	4593
Estables, village, Haute Loire	Granite	,,	4436
Road from Le Puy to Pradelles, Haute Loire	Basalt on granite	,,	4203
Puy de Laschamp, Monts Dôme	Scoriæ	*Ramond*	4196
Petit Puy de Dôme, Monts Dôme	Scoriæ	,,	4187
Puy de Côme, Monts Dôme	Scoriæ	,,	4173
Puy de Pariou, Monts Dôme	Scoriæ	,,	4012
Puy de Cliersou, Monts Dôme	Trachyte	,,	3992
Puy du Petit Suchet, Monts Dôme	Scoriæ	,,	3983
Puy de Louchadière, Monts Dôme	Scoriæ	,,	3956
Puy de Las Solas, Monts Dôme	Scoriæ	,,	3950
Lac Paven, Monts Dore	Basalt	,,	3943
Puy de Mont-char, Monts Dôme	Scoriæ	,,	3933
Puy de Chopine, Monts Dôme	Trachyte	,,	3910

TABLE OF HEIGHTS.

			Eng. feet above the Sea.
Puy de Jume, Monts Dôme	Scoriæ	*Ramond*	3848
Pey-Veuy, Haute Loire, near Montbonnet	Scoriæ	*B. Roux*	3822
Montagne de Bar, Haute Laire	Scoriæ	,,	3816
Grand Sarcoui, Monts Dôme	Trachyte	*Ramond*	3799
Puy des Goules, Monts Dôme	Scoriæ	,,	3796
Puy des Montgy, Monts Dôme	Scoriæ	,,	3792
Tête de la Serre, Monts Dôme	Basalt on granite	,,	3461
Puy de Chatrat, Monts Dôme	Basalt on granite	,,	3372
Puy de Manson, Monts Dôme	Basalt on granite	,,	3310
Puy de St. Sandoux, Monts Dôme	Basalt on granite	,,	2821
Puy Girou, Monts Dôme	Basalt on fresh-water marly limestone.	,,	2792
Puy Barnère, Monts Dôme	Basalt on fresh-water marly limestone.	,,	2780
Orcines, village, Monts Dôme	Granite	,,	2779
Graveneire, Monts Dôme	Scoriæ	,,	2723
Prudelles, Monts Dôme	Basalt on granite	,,	2313
Côtes de Clermont, Monts Dôme	Basalt on fresh-water marly limestone.	,,	2089
Puy de Dallet, near Pont du Château	Basalt on fresh-water marly limestone.	,,	2007
Threshold of Hotel de Ville at Le Puy		*B. Roux*	1710
Puy de Crouel, Limagne	Calc peperino	,,	1430
Clermont, Place de Jaude, Puy de Dôme		,,	1286
Level of the Allier at Pont du Château		,,	1026

EXPLANATION

OF THE

MAPS AND ENGRAVINGS

ILLUSTRATIVE OF THIS MEMOIR.

MAPS.

No. 1.

"The Volcanic District of Central France."

Comprising the greater part of the primary platform, and containing the departments *Puy de Dôme, Loire, Rhône, Haute Loire, Cantal, Ardèche, Lozère,* and parts of the *Aveyron, Corrèze, Creuse,* and *Allier.* The areas occupied by the volcanic products are distinguished in colours, as well as those of the principal crystalline and sedimentary formations, and the several coal-basins. This map is copied for the most part from the 'Carte Géographique de la France,' omitting the subdivisions, which would have been too complicated for the scale employed.

No. 2.

"The Monts Dôme and Part of the Limagne."

Exhibiting the chain of puys rising from the granitic platform which separates the Allier and Sioule. The extent of surface covered by the volcanic cones and their lava-currents is indicated by colours, as well as the limits of the primary and freshwater formations, and the more ancient basaltic currents which have flowed from the Mont Dore in that direction.

29 AP 58

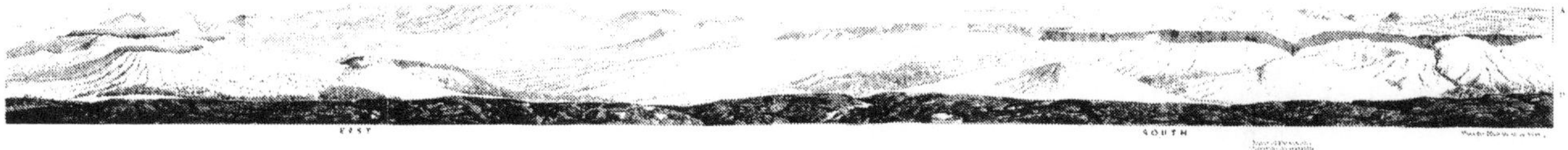
EAST
SOUTH

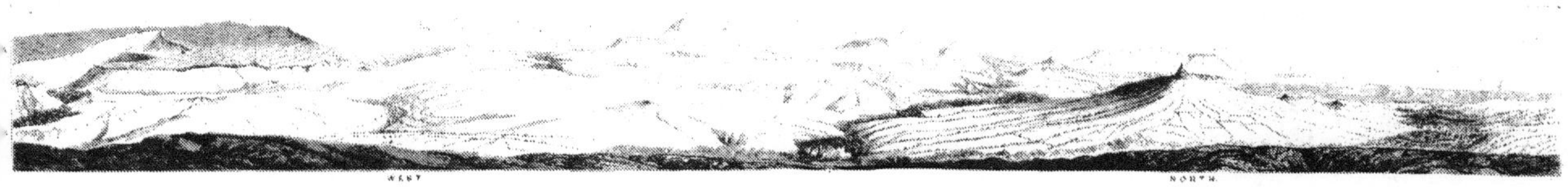
WEST
NORTH

Plate I.

Panoramic View of the Environs of Clermont, taken from the summit of the Puy Girou, a conical peak of columnar basalt, about four miles south of that town. The chain of puys of the Monts Dôme is seen to the west, with the Puy de Dôme in the centre, rising from the surface of the granitic platform. To the north lies the basin of Clermont, eaten out of the freshwater formation, and overlooked by the primary escarpment, on the edge of which rises the recent volcanic cone of Graveneire. Beyond Clermont are seen the basaltic platforms of Les Côtes and Chanturgues; in front, a basaltic peak crowned by the Castle of Montrognon. Eastwards, and above the village of Opme, rise the Puy de Jussat and the plateau of Gergovia, both in all probability once continuous with the basalt of Girou. To the south is seen in its whole extent the basaltic platform of La Serre, partly resting on granite, partly on the freshwater marls, and terminating at the village of Le Crest. Beyond it rise the trachytic heights of the Mont Dore, and its basaltic embranchments; with the insulated basaltic plateaux or peaks of St. Sandoux, St. Saturnin, Coran, Monton, &c.

The horizon is closed by the mountains of the department of the Haute Loire. Immediately below these distant heights the valley of the Allier appears narrowed by the quantity of basalt which has flowed in that direction from the vicinity of the Mont Dore. Beyond the river rises an insulated group of hills capped by basalt and chiefly composed of freshwater limestone, but partly of granite and secondary sandstone. They are called the Puys de Millefleur, St. Romain, Mauriac, Dallet, &c. At the foot of this range flows the Allier, and immediately enters the wider expanse of the Limagne d'Auvergne, which, bounded only eastwards by the distant granite range of the Forèz, stretches to the horizon on the north-east of the station. The cone of Graveneire, and the lava-current bursting from its side and spreading thence over a large surface of the lower plain, are remarkable objects to the north of the spectator.

Plate II.

Distant View of the Chain of Puys, or Monts Dôme, taken from a remnant of basalt crowning the western bank of the Sioule, a short distance below Pont Gibaud. Nearly all the cones are visible; the most conspicuous are those of Louchadière and Côme. The lava-streams poured forth by these two vents may be seen spreading in broad sheets from the base of the cones over a wide extent of the granitic platform, and uniting immediately above the Castle of Pont Gibaud to pour over the banks of the valley of the Sioule in a spreading sheet, which is, however, partly hidden by forest and underwood. That part of the lava-current which occupies the former bed of the Sioule is, on the contrary, still bare and rugged, its surface appearing like a series of heaps of loose basaltic blocks. Between this and the granite cliffs to the right, the river dashes and foams through the narrow channel it has excavated by undermining the granite. At the angle immediately beneath the spectator the basalt exhibits a regularly columnar structure.

In the distance to the right is the outline of the central trachytic eminences and plateaux of the Mont Dore, from amongst which the Sioule takes its rise.

Plate III.

Transversal View of the Northern Chain of Puys, from the Summit of the Puy Chopine. The cone of La Goutte, which half encircles the Puy Chopine, forms the foreground. The Puy de Côme and the commencement of its lava-current are seen to the extreme right. The Puys de Dôme, Cliersou, Suchets, Pariou, with the beginning of its lava-stream, &c., in the front; and the group of Les Goules and the Sarcouis, with the Puy Chaumont, on the left. The Mont Dore and the mountains of the Forèz and Haute Loire, skirt the horizon to the right and left. The observer looks due south.

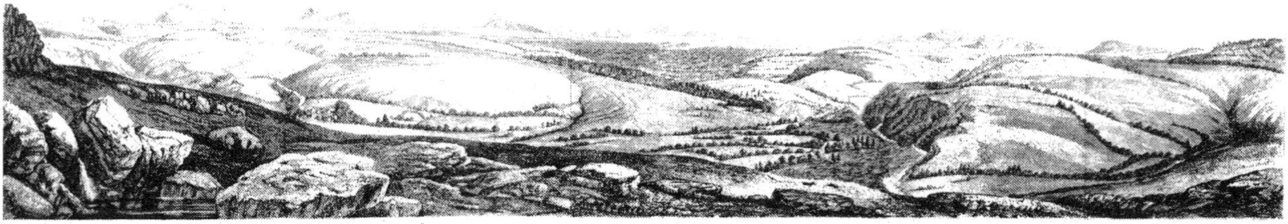

29 AP 58

TRANSVERSAL VIEW OF THE MONTS DOME FROM THE SUMMIT OF THE PUY CHOPINE.

Plate III.

29 AP 58

Plate IV.

Lateral View of the Northern Chain of Puys from the east, taken between Volvic and Channat. The cone and crater of Nugère are observable on the right; a current of lava appears still to boil over on two of its sides; after encircling a knoll of granite it enters the valley of Volvic, which it threads in its course towards the plain of the Limagne. Immediately to the left of the Puy Nugère is the wooded group of cones called the Puys de Jumes, la Coquille, and Leironne; from the foot of which a sheet of lava overspreads the surface of the granite platform, and, entering a narrow gorge between the point of view and the Puy Channat on the left, finds its way into the low plain of the Limagne, which it has flooded to a considerable extent. (See the Map of the Monts Dôme, and p. 77 et seq.) The figure of the Puy du Grand Sarcoui as seen from this point is very remarkable. It is one mass of porous and earthy trachyte, and rises between and in close contact with two regular cones of scoriæ, the Petit Sarcoui and des Goules. The Puy Chopine is observable in the middle of the sketch,—a strange mass of rock, half granite, half trachyte, resting upon basalt, and rising out of the crater of the Puy de la Goutte.

Plate V.

Transversal View of the Southern Chain of Puys, taken from the Summit of the Puy de la Rodde. In the middle appear the broken-down craters of Chaumont, Las Solas, and La Vache; from the bottom of each a lava-current is seen to issue. Those of the two last cones descend together to the east, passing behind the Puy Chaumont, and have formed the lake of Aidat by damming up a river called Le Veyre, which flows from the Mont Dore. Below the lake this lava continues to occupy the narrow channel of the river, excavated through granite as far as the village of Talande, a distance of fifteen miles. (See the Map.)

Beyond the lake are seen the basaltic platform of La Serre, and the Castle of Montredon, built upon an insulated peak of basalt, once continuous with the former. To the right are the Puys Combe-grasse, La Taupe, Montgy, Pourcharet, Montjughat, and Monchal. At the foot of the last stands the farm of *M. le Comte de Monlosier*, author of 'Essai sur les Volcans d'Auvergne,' who established himself in the midst of this volcanic desert, and succeeded in turning its arid plain of scoriæ into productive corn-land; for, notwithstanding the great elevation of the spot, it is completely sheltered by the encircling range of volcanic cones, and the fertility of the soil is unquestionable. Here he died in 1840, and here was buried; the influence of the Jesuits, whose intrigues he unmasked in his admirable political writings, having prevented his interment in any consecrated burial-ground.

Plate VI.

Valley of Villar, and Plateau de Prudelle on the summit of the granite slope to the left. Below a Roman paved road is based on the comparatively recent lava-current of Pariou. The ravine to the right has been excavated since the flow of this lava through the granite bank to the right.

Plate VII.

This general sketch of the Mont Dore is taken from the foot of the Puy Gros, whence a bird's-eye view is obtained of the Valley of the Dordogne, in which lies the village of the Baths. At its upper extremity is the deep circus which appears to have been the site of the chief vent of the volcano. The principal peaks which encircle this gorge are named in the references. From those called the Roc de Cuzau on the left, and Le Cliergue on the right, wide platforms of trachyte slope gradually towards the observer. The first form the Plateaux de Durbise and de Langle; they may be seen in the sections afforded by the valley of the Baths to rest upon tufa, this again on a parallel bed of clinkstone, and this on basalt with

29 AP 58

Plate IV

the interposition of conglomerate. At the Cascade du Mont Dore, immediately above the Baths, this superposition is very evident. The Plateau de Rigolet, on the other side of the valley, is similarly constituted, as may be seen in the valley of La Scie on its opposite side. By this it is separated from another extensive embranchment consisting of repeated beds of trachyte, and forming the Plateau de Bozat, from beneath which basalt crops out at the Plaine de Chamablanc. The similarity in height, slope, and structure of the Plateaux de Langle and Rigolet leads to the opinion that they were once continuous. Their separation may have been effected by a local earthquake, and enlarged by the erosion of the river. The left of the field of view is occupied by a series of swelling trachytic eminences, of which it is difficult to determine whether they were thrown up in their present irregular and rounded forms, like the trachytic puys of the Monts Dôme, or produced by the unequal weathering of a massive hummock. Basalt appears to crop out also from beneath these trachytes on the north-east side; but from the quantity of débris of trachyte accumulated along the lines of junction, it is difficult to find any positive fact of superposition. The Puys de Loueire and Trioulerou, at the extreme left, are of very slaty trachyte (clinkstone), a variety which predominates in that direction.

Plate VIII.

View towards the north, from near the Puy Gros, looking down one of the tributary valleys of the Sioule, towards Rochefort. The rocks Sanadoire and Tuilière (clinkstone) stand conspicuous on either side, and the Puy de Loueire on the extreme right.

Plate IX.

A View of the Valley of Chambon, one of the principal excavations which intersect the Mont Dore. It has its source in the centre of the group, and is directed in a straight line towards the east. The deep gorge called Chaudefour, which forms its upper extremity, probably one of the last central craters of the volcano, is bounded by perpendicular and degraded cliffs of trachyte, offering three or four varieties, superposed to one another in irregular beds with intervening layers of tufa, the whole penetrated by dykes of trachytic rock. At no great distance from the central heights thick beds of basalt overlie first the trachyte, afterwards the fundamental granite, and, spreading far and wide into extensive plateaux, descend towards the main valley of the Allier with a gradual slope, which is, however, broken on some points by subsequent excavations, on others, to all appearance, by partial elevations, occasioned perhaps by some of the convulsive throes of the volcano. These sheets of basalt are accompanied throughout their whole extent by irregular deposits of volcanic conglomerate, varying from a coarse basaltic breccia to a fine pumiceous tufa. The conglomerates generally support or alternate with the basalt, particularly towards the lower part of the valley, where the granite is entirely concealed by the accumulated volcanic products. The chief interest of this view lies in the recent volcanic vent of Tartaret, which has exploded in the middle of the valley, and by the quantity of matter it ejected dammed up the river so as to form the Lake of Chambon, which is easily seen to have extended originally up to the entrance of the Vallée de Chaudefour, but which has been lowered by the wearing away of its canal of discharge. Immediately to the left of the cone of Tartaret may be perceived the commencement of its lava-stream, which continues to occupy the former bed of the river for a distance of thirteen miles. Further to the left rises the ruined Château de Murol, once a very strong fortress belonging to the family of d'Estaing.

Plate X.

Montagne de Bonnevie, a colossal cluster of columnar basalt, above the town of Murat (Cantal). A detached segment from the repeated currents of basalt which have flowed from the central heights, and are seen behind in cliff ranges.

Plate VII.

Plate IX.

29 AP 58

Plate XI.

Panoramic Sketch of the Basin of Le Puy and the Mont Mezen, taken from the Mont d'Ours near that town (Haute Loire).

The view from this point is exceedingly instructive, embracing all the principal features of an interesting and singular country. The Mont d'Ours is a double volcanic cone of recent formation, although its craters have disappeared. The eruptions by which it was produced burst through a vast bed of volcanic breccia and basalt, which is still visible at the village of Ours, and forms a succession of rocky terraces from thence towards Le Puy. It rises near the eastern limit of the freshwater formation, the remains of which are seen on every side, resting against the primitive margin of the cavity in which it was deposited. To the south-east are seen the highest summits of the Mezen; and from thence to the north the horizon is bounded by a chain of rocky eminences, the ruins of an enormous current of clinkstone, which that volcano has poured forth towards the north. (See p. 157.) The last of these rocks, called *Miaune*, and another not seen in the drawing, near the town of Roche en Regnier, rise from a base of granite on the *further* side of the Loire; the current appearing to have occupied the former bed of that river to some distance, and to have forced it to excavate a fresh one through the whole mass of phonolite. The basaltic currents of the Mezen, which, accompanied by prodigious eluvial accumulations of brecciform conglomerate, overwhelmed the whole surface of the freshwater formation, are observable in all directions; and it is evident, from their being found on both sides of the Loire, that this river's actual bed was not yet formed at the æra of their descent. The channel by which the Loire now issues from the basin of Le Puy may be observed to have been pierced entirely through granite, although at the very edge of the calcareous formation;—a remarkable fact, seeming to show that even the softest marls under a protecting cover of basalt resist meteoric erosion far more effectually than granite itself.

It was by the excavation of this narrow, deep, and tortuous gorge, that the waters dammed back in the basin of Le Puy could only be discharged. As the channel gradually deepened, the valleys that now intersect its tertiary and volcanic formations were progressively formed. That this was effected by degrees, and not by any sudden debacle, is proved, 1st, By the undisturbed state of the cones of scoriæ. 2ndly, By the great variety of levels at which the basaltic lava-beds occur: they appear, in fact, as in the valley of the Limagne, to be the more recent the lower the level they occupy, thus marking the successive steps of the process of excavation. (See p. 180.) It is, however, possible that some cleft broken through the granitic barrier by subterranean expansion may have assisted the discharge of the lake in that particular direction.

The prospect to the west is equally interesting. The horizon on that side is limited by the vast chain of volcanic cones, which, commencing at the foot of the primitive heights of La Chaise-Dieu, reaches uninterruptedly to Pradelles. The lava-streams produced by these eruptions have deluged an extensive and nearly level plain, stretching from the foot of the range to Le Puy. This is intersected by the channels of various torrents, and may be seen to consist of repeated currents of basalt, resting on the freshwater limestone. A few insulated cones are scattered over the plain, and some occur at a considerable distance from the line upon which the greater number of eruptions have broke forth. That near the village of Chaspinhac, on the slope of a massive granitic embranchment rising from the river Sumène, is remarkable for its elevated position. It has apparently furnished the lava-current of the Plateau de Fay, which, however, is now separated from it by the deep gorge by which the Sumène joins the Loire. The large cone of Bar, near Alègre, which still presents a very perfect crater, and those of St. Geneys and Couran, are conspicuous in the distance; as well as those in the opposite direction which rise from the Forest of Bréysse, near Prézailles, on the western slope of the Mezen.

The valley of the river Borne, from which rises the town of Le Puy, the capital of the department of the Haute Loire, is not the least interesting part of the view. Its numerous feeders, which unite near the remarkable insulated and regularly columnar rock called "The Organ of Expailly," have each their channel cut through a system of basaltic beds, generally columniform, often double, with a bed of conglomerate or alluvial matter interposed, and resting on the freshwater limestone strata. The great mass of volcanic breccia called the Rocher Corneille is seen rising above Le Puy, as well as the spiry pinnacle of St. Michel, formed of the same substance, and standing up in the middle of the town. A similar mass rises from the town of Polignac, as seen above, at a distance of about four miles. These rocks were favourable sites, of course, for châteaux-forts in the *good old* times of rapine and butchery; and most of the towns of the central provinces of France were built under the protection of some such fortress.

There are perhaps few spots on the globe which offer a more extraordinary prospect than this. To the eye of a geologist it is superlatively interesting, exhibiting in one view a vast theatre of volcanic formation, containing igneous products of various nature, belonging to different epochs, and exhibited under a great diversity of aspect.

Plate XII.

View of the Extremities of the south-western lateral Embranchments from the Basaltic Platform of the Coiron, as seen from the neighbourhood of St. Jean le Noir, on the road from Villeneuve de Berg to Viviers (Ardèche).

Plate XIII.

Volcanic Cone and Current of Jaujac (Ardèche). The cone rises from the middle of a trough-shaped valley, occupied by a coal-formation, and bounded on either side by ranges of granite and gneiss. A lava-current may be seen to proceed from a breach in the circuit of the crater, and descend into the valley of the Alignon, which it once filled to a considerable depth and width. The excavation since effected by the river has exposed a mural range of columnar basalt about 100 feet high, and extending uninterruptedly into the further valley of the Ardèche, and even some distance down the course of the latter river. On the edge of this cliff-range stands the village of Jaujac. In the distance to the left are seen the primitive heights of the Haut Vivarais; and through the opening afforded by the valley of Prades a part of the sloping basaltic platform of the Coiron. The group of hills that rise in the centre of the view are of granite. The summit of another neighbouring volcanic cone, called La Gravenne de Soulliols, is just visible on the left.

Plate XIV.

The Valley of Montpézat. This view is taken near the confluence of the torrents Fontaulier and Pourseille, whose channels are eaten out of a massive bed of basalt produced by the volcanic cone which rises to the extreme left. The platform dividing them terminates in an exceedingly acute angle, composed of very beautiful and regular columns resting on granite. A ruined castle stands on the point of this angle, and adds its picturesque effect to the natural beauties of this very lovely scene. The valley is encased by the granite buttresses of the platform of the Haut Vivarais. Their skirts are clothed with rich forests of Spanish chesnut. In the distance rises a volcanic cone of inferior size to that of Montpézat, and which has not produced any lava-stream.

It may be remarked that the vast amount of excavation which has taken place in this valley since the epoch at which it was filled up to the level of the Castle of Pourchirol with liquid basalt, and this not only through the basalt but to a depth of more than 100 feet through the underlying granite (which is of a very hard compact nature), *can* only have been effected by the erosive force of *the torrents that flow there at present;* since any more general and violent rush of waters, of the nature of a deluge or debacle, would not have left undisturbed the two cones of loose scoriæ, of which parts are seen in our sketch, and of which the materials are so incoherent that the foot sinks ankle-deep into them.

29 AP 58

SECTION OF GRANITIC PLATEAU FROM N.N.E. TO S.S.W.

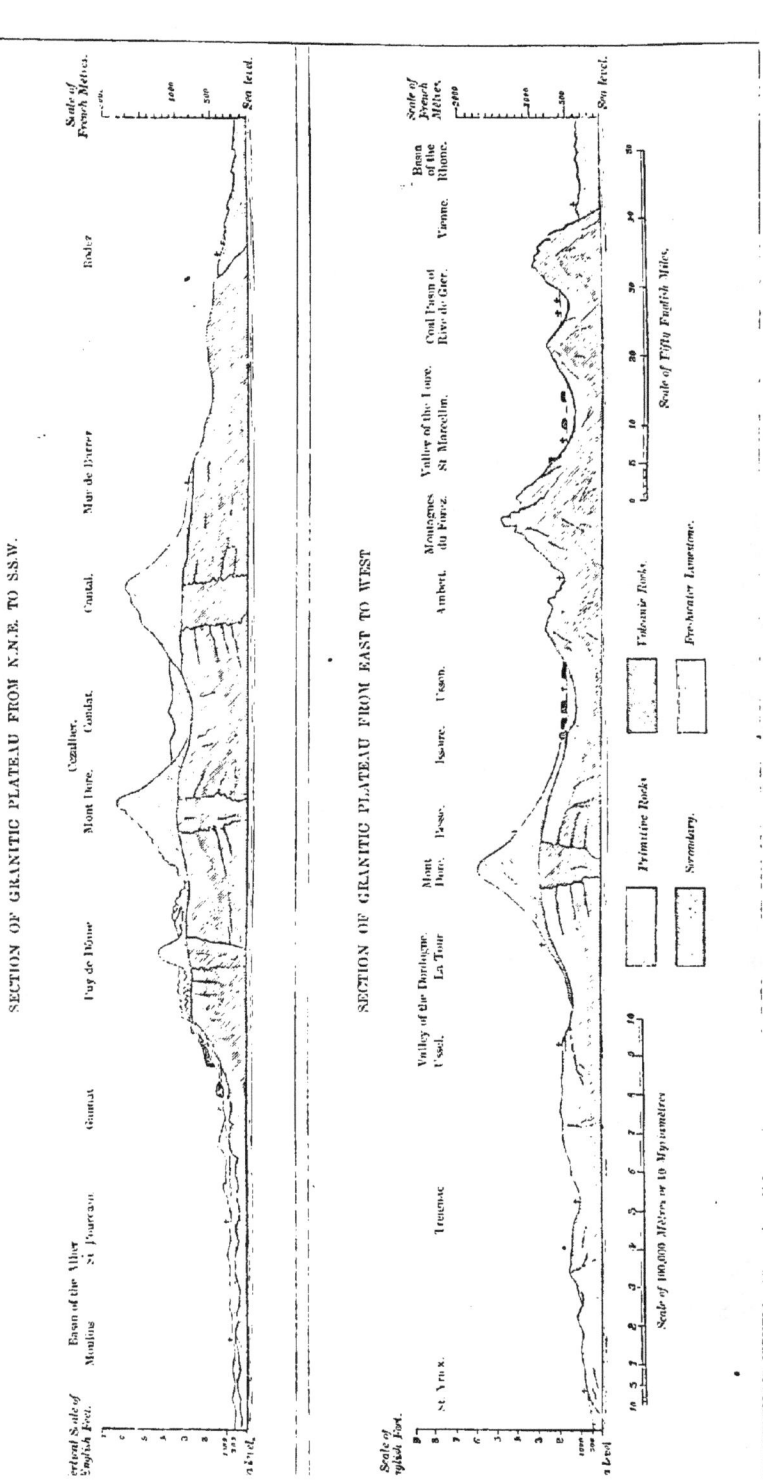

SECTION OF GRANITIC PLATEAU FROM EAST TO WEST

Plate XVI.

Plate XV.

La Coupe d'Ayzac, a volcanic cone with a very perfect crater, near Antraigues (Ardèche). The crater being on the opposite side of the hill is not seen in the drawing.

A lava-current descending from it has been cut through by the river Volant. Its section forms a vertical rock, exhibiting three apparently distinct ranges of columns. Those of the lower story are very regular. The basalt rests on either side against the granitic cheeks of the valley whose bottom it occupied. This drawing was taken from the outskirts of the village of Antraigues, and from the foot of a remarkably isolated rock of columnar basalt, lying precisely at the same level with those seen on the opposite side of the river, with which it was certainly once continuously united. Similar but larger patches of the same current of basalt are visible on many points down the valley of the Volant for a distance of four or five miles, particularly in its concave angles. The excavation to which their insulated position is owing suggests the same reflection as that seen in the preceding engraving of the valley of Montpézat. (See p. 192.) By comparing this drawing with that given by Faujas of the same cone, it will be seen that in his directions to his draughtsman he must have trusted to memory, and that a very bad one. Besides the complete dissimilarity of form, it is absolutely impossible to see the crater of La Coupe and the basaltic range of columns above the river, together, from any one spot.

Plate XVI.

Sections of the Primitive Platform of Central France from N.N.E. to S.S.W. and from east to west.

Plate XVII.

Fig. 1. Composite Sections or Profiles of the principal Hills in the immediate neighbourhood of Clermont, showing the various heights at which contiguous beds of basalt are now found in the basin of the Limagne, tending to prove the gradual excavation of that valley. (See p. 202.)

Fig. 2. Section of the Mont Mezen and part of the Freshwater Formation of Le Puy (Haute Loire). This section is taken from M. Bertrand Roux's excellent work on the Environs of Le Puy. The map which also accompanies that volume will be found very serviceable to geologists who may wish to explore that interesting district themselves.

N.B. In both these sections the vertical scale is of course greatly magnified as compared with the horizontal.

INDEX.

ÆTNA.

ÆTNA, 30, 39, 42, 43 n., 44, 134 n.; compared with Mont Dore, 115, 116 n.; great extent of some of its lava-currents, 117 n.; luxuriant growth of its chesnut-trees, 190.
Agde, 196.
Aidat, lake of, 74, 92, 103.
Aigueperse, 12.
Aiguiller, Puy de l', 126, 131, 137, 138.
Aix, 196.
Alagnon, the, 146, 150, 151.
Alais (Gard), 3, 4.
Alassac (Corrèze), 3.
Alègre, 155, 184.
Alignon, the, 190, 191.
Allier, 2, 4, 7, 15, 17, 20, 26, 40, 112, 134, 135, 139, 146, 154, 155, 176, 177, 184, 211; probable cause of the fall of its celebrated bridge, 154 n.; extreme violence of its floods, ib.
Alpes, Hautes, 1.
Alps, the, 186, 212.
America, aqueous phenomena of the trachytic volcanos of, 122, 123 n.
Amethysts of Le Vernets, 2.
Andes, the, 134 n.
Angle, Plateau and Puy de l', 123, 131, 132, 133.
Antraigues, 245.
Anza, 140.
Apcher, 140.
Apennines, the, 186 n.
Ardèche, 3, 4, 38, 39, 154, 155, 162, 166, 175, 187, 188, 189, 191, 192, 194, 195.
Ardes (Puy de Dôme), 3, 140.
Arniat, 77.
Artias, 186, 193.
"Arkose," M. Brongniart's application of the term, 8.
Arverni, traditional site of the city of the, 106 and n. †.
Asia Minor, analogy between the volcanic rocks of, and those of Central France, 210 n.
Aubenas, 38, 155, 162, 176, 185, 192; description of the basaltic dyke in its vicinity, 195, 196.
Aubepin, 169.

BASALT.

Aubières, 108.
Aubin (Aveyron), 4.
Aubrac, Canton d', volcanic region of the, 152, 153.
Aulière, 187.
Aumône, Cheire de l', 55, 58, 61.
Aurillac, 24, 25, 132, 148, 200; auriferous origin of its name of, 151 n.
Autun (Saône et Loire), 4.
Auvergnat, 79.
Auvergne, 1, 2, 7, 11, 26 n., 28, 31, 32, 33, 34, 38, 43, 47, 58, 60, 72, 81, 96 n., 100, 169, 176, 178 n., 181, 198; number of ruined "châteaux-forts" on its peaks, and ultimate fate of their lawless possessors, 102 n.
Auvergne, la Tour d', 139.
Auvergne, Limagne d'. See Limagne.
Auvergne and Forèz, surface elevation attained in the regions of, 1; vestiges of deposits of freshwater lakes there, 2.
Aveyron, 3, 4, 5, 152.
Aymard, M., 27, 35 n., 178, 218; his arrangement of the fossils, &c., of the Haute Loire district, 229-231.
Ayzac, la Coupe d', 186, 193; beauty of its crater and basaltic columns, 193, 194

Baladon, Puy de, 133.
Bannière, Puy de la, 44, 78, 82, 83.
Banson, Puy de, 87.
Bar, Montagne de, 184; its height, 233.
Baraque, La, 63.
Barbier, Puy de, 133.
Barbu, le Roc, 127.
Barme, Puy de, 88.
Barmet, Puy de, 61.
Barnère, Puy, 211; its composition and height, 233.
Basalt of Chalucet, composition of the, 98; preservative agency of the disposition of basaltic prisms, 106 n.; olivine not invariably a characteristic of basalt, 130 n.; question of the alternate superposition of trachyte and basalt, 131 n.; favourable occasion for observing the igneous origin of basalt, 142 and n.; interesting character of the basaltic

BASALT.

prisms of the Montagne de Bonnevie, 150; composition and formation of basaltic dykes, 173-175.
Basalt of the Haute Loire and Ardèche (Velay and Vivarais). *See* Aubenas, Ayzac, Burzet, Coiron, Jaujac, Mont Mezen, Montpézat, Souillols, Thueyts, Velay, Vivarais.
Bassignac, 4.
Baume, La, 191.
Bauzon, forest of, 185.
Baveno, 2.
Beaune, Valley of, 89.
Beaumy, Puy et Lac de, 79, 80, 82.
Beauregard, 19.
Beauzac, 157.
Bedarrieux (Herault), 4.
Benoit, Puy, 112.
Bert (Allier), 4.
Berzé, Puy de, 106; its site, 108; its altitude compared with that of adjacent puys, 109; this puy the probable source of the basaltic current of Gergovia, 204.
Besace, Puy de, 88.
Besse, Valley of, 134, 143.
Beudant, M., remarks on the views of, relative to the superposition of trachyte, 131 and 132 *notes*, 147.
Beurdouze, 144.
Béziers, 196.
Blaves, 178.
Boën, 28.
Boiseghoux, 100.
Bone-beds of Mont Perrier, locality of the, 135.
Bones. *See* Human Bones.
Bonnevie, Montagne de, 150.
Borne, the, 175, 177, 181, 183 *n*.
Bort (Puy de Dôme), 4, 149.
Bouchet, Lake du, 185.
Bouillet, M., 29 *n*., 35 *n*., 87 *n*., 217.
Bourboule, la, 23.
Bourglastic, 4.
Bouttières, les, 167.
Bozat, Plateau de, 132, 133.
Brassac, 4.
Bravard, M., 135, 217.
Breccia, human bones found in, 181, 182.
Breislak, M., 48 *n*., 120 *n*., 122 *n*.
Bridges, natural, formed by incrusting springs, 22; single-arch bridge over the Allier, 154 *n*.
Brion, 144.
Brioude, 7, 154, 182, 211.
Brives (Corrèze), 4.
Brocchi, M., 19 *n*., 48 *n*., 120 *n*.
Brongniart, M., 8, 11.
Brousson, Puy de, 94.
Brunelet, 173.

CAPUCIN.

Buch, M. von, 33, 48, 51, 96 *n*., 180 *n*.
Building-stone, quarries of, 78, 128.
Burat, M., 35 *n*.
Buron, Pic de, 113.
Burzet, volcano of, 186, 187, 192; frequency of olivine nodules in its basaltic bed, 188; illustration of contractile force afforded by these nodules, *ib*. and *n*.; idea suggested to the natives by the horizontal sections of basaltic columns, 188, 189; form and dimensions of the columns, 189.
Butte de Montpensier, 12.

Cacadogne, 125, 126, 133.
Caddis-fly. *See* Phryganea.
Cæsar. *See* Julius Cæsar.
Caissière, La, 92.
Calcareo-volcanic strata of the Limagne, components and characteristics of the, 15-17.
Calcareous peperino, examples of, where found, 18-21; analogy between those of the Limagne and of the Vincentin, 19 *n*.; beauty of that of Pont du Château, 20.
Calcariferous springs of Auvergne, localities of the, 21-23.
Camaldoli, 78.
Canary Isles. *See* Lancerote.
Cane, Grotta del, 85, 193.
CANTAL, 3, 38, 39, 48, 49, 114, 115, 124 *n*., 127 *n*., 134 *n*., 139, 140, 144, 155, 156, 167, 172, 175, 198, 199, 200, 209, 211; vestiges of deposits of freshwater lakes in the, 2; similarity of its freshwater formation to that of the Limagne, 24; its locality and distinctive features, *ib*.; difficulty of ascertaining its original limits, 25.
VOLCANIC REGION OF THE CANTAL, its figure: point of difference between its lava-current and that of the Mont Dore, 145; character of its valleys, 145, 146; bulk, extent, and character of its conglomerates, 146; their origin due to combined aqueous and igneous action, 147; site of its central crater, 148; vast extent of its basaltic beds, 149; interesting configuration of the basaltic prisms of the Montagne de Bonnevie, 150; difficulty in removing them unbroken, *ib*.; alleged former auriferousness of the Jourdanne river-sand, 151 *n*.; relative age of the Mont Dore and Cantal volcanic remains, not determinable, 152; cause of impediment to the basaltic ramifications E. and S.E. of the latter region, 154; height of the Plomb du Cantal, 232.
Capucin, le, 131.

INDEX.

Carladez. *See* Vic en Carladez.
Cassini's map, 52.
Catania, 65, 157 *n*.
"Causses" of the Cevennes, 4.
Central France. *See* France, Central.
Cèr (or Cère), the, 25, 146, 148, 151, 152.
Cevennes, elevation of the rocks in the region of the, 2; defensive advantages of the district to the Protestants persecuted by Louis XIV., 4, 5.
Cezallier, Montagnes de, 139.
Chadrat, 10.
Chaise Dieu, la, 26, 154, 157.
Chalar (or Chalard), Puy de, 80, 81, 94.
Châlons sur Saône, central position of, 1.
Chalucet, 23, 49, 198; composition of its basalt, 98.
Chamablanc, 132.
Chamalières, 26, 179.
Chambeyrac, 178.
Chambon, 114, 124, 133, 134, 135, 140, 141.
Champagne, 197.
Champanelle. *See* St. Genest de Champanelle.
Champedaze, 144.
Champeix, 141.
Chandefour, 124.
Channat, Puy, 44, 87, 104, 112; no vestiges of its crater visible, why? 83.
Channonat, 108, 204.
Chantouzet, 137.
Chanturgue and Clermont, Côtes de, once evidently united though now two, 104.
Charade, Puy de, 84, 98; reasons for assigning a recent origin to it, 99, 100.
Charlanne, 132.
Charmont, Puy de, 94.
Chaspinhac, 178, 179, 184.
Chastel, 150.
Chastriex, 134 *n*.; 139.
Château-gay, points of similarity between the plateau of, and that of Jumes, 100, 104.
Châteauroux, central position of, 1; its lithographic limestone, 13.
Chateix, 98.
Châtelguyon, 23.
Chatrat, Puy de, 109; its height, 233.
Chaude de Fay, 178.
Chaudesaigues, 23.
Chaumont, Puy de, 71, 76.
Chauriat, 19.
Chauvet, 144.
"Cheires," rugged fields of lava, 43; etymology of the term, *ib. n.*; cause of their ruggedness, 92. *See* Aumône, Côme, Pariou.
Cheix, les, 86.

Chesnut-trees, volcanic soil most favourable to the growth of, 190.
Chidrac, 135.
Chiliac, 184.
Chimborazo, 69.
Chopine, Puy, 46, 71; a perplexing enigma to the geologist, 72; the author's deductions regarding it, 73–75; origin of its elevation, 75; its height, 232.
Chuquet Geneto, 87; explanation of the term "Chuquet," *ib. n.*
Civita Vecchia, 120.
Clermont, 20, 21, 22, 23, 31, 33, 34, 40, 67, 85, 86, 98, 132, 211.
Clermont and Chanturgue, Côtes de, once evidently united, though now separated, 104.
Clermont Ferrand, 39.
Clersat, 66.
Cliergue, Le, 128, 132,
Cliersou, Puy de, situation and peculiar shape of the, 66; its caves perforated by the Romans, *ib.*; M. Ramond's hypothesis relative to this puy, 67; its height, 232.
Clinkstone of Mont Mezen. *See* Mont Mezen.
Coal-measures, strata associated with the, 3; localities of the principal coalfields, 4.
Coiron, volcanic region of the, 161, 162; fanciful appellation bestowed by the peasantry on its basaltic phenomena, 163, 164; basaltic rock of Rochemaure, 164; instructive features connected with the basaltic currents of this region, 165, 166.
Col de Cabre, 149; its height, 232.
Colière, Puy de, not noticed by previous writers, 86; characteristics of its basaltic products, 86, 87.
Combegrasse, Puy de, 94.
Côme, Puy de, regularity of the conical form of, 55, 56; prodigious dimensions of its lava-current and phenomena connected therewith, 56–61; its height, 232. *See* also 66, 79, 89.
Compains, Valley of, 143.
"Cones of eruption," 41 *n*. †.
Conglomerates of Mont Dore, *see* Mont Dore; tendency of conglomerates to waste into pyramidal form, 172 *n.*
Coquille, Puy de la, 77.
Coran, Chox de, dimensions of the crater, and composition of the strata of the, 111.
Coran, Puy de, 13; circumstances which entitle it to attention, 111, 112.
Cordier, M., 35, 232.
Cordilleras, the, 134 *n.*
Corneille, Rocher, 172, 175, 181.
Cornon, Puy de, once the bed of the Allier, 112.

CORRÈZE.

Corrèze, 3, 4.
Côtes, Les, near Clermont, 12, 64; their height, 233.
Coteuge, 144.
Cour, Vallée de la, 126, 132.
Cour-cour and Montagne de Cour-cour, 18, 19.
Couron, 184.
Craters, intenability of the theory of "elevation craters," 84 n.; remarks on the alternating conditions of craters, 127 n.
Crest. *See* Le Crest.
Creux Morel, le, and its singular crater, 70, 71.
Creuze (Herault), 4.
Crimea, the, 122.
Croix de la Paille, 181, 182.
Croix des Bouttières, 156, 167.
Croix Morand Plateau, 232.
Croizet, M., 26 n., 135.
Cros de Pèze, 137.
Crouel, Puy de, 18, 21; its height, 233.
Croustaix, Montagne de, 180.
Crouze, the, 23.
Crystals of certain localities of Central France, beauty and variety of the, 2, 20, 21.
Cuzau, le Roc de, 126, 128, 133.

Dallet, Puy de, 12, 15, 17, 112, 200; its height, 233.
Daubeny, Dr., 35, 37.
D'Aubuisson, M., 34, 51, 65, 72, 122 n.
Davayat, 12.
Decize (Nièvre), 4, 211.
Denise, Montagne de, 180, 181, 182, 183.
Dent du Marais, 135.
Desmarest, M., 31, 32, 33, 47, 52 n.
Dienne, 149.
Diluvian theory, remarks on the intenability of the, 204, 207 n.
Dogne, the, 116, 125. *See* Dore and Dogne.
Dolomieu, M., references to the opinions of, 32, 48.
Dôme, Monts. *See* Puys, chain of.
Dôme, Petit Puy de. *See* Puy, Petit.
Dôme, Puy de, elevation, &c., of the, 45; composed entirely of *domite*, 45-47 (*see* Domite); its measurement, &c., 52, 53, 232; mode of its production not indicated by its structure, 54. *See* also pp. 3, 4, 39, 41, 48, 49, 50, 51, 66, 68, 69, 78, 86, 87, 89, 129, 155, 232.
Domite, a variety of trachyte, why so called, 45; its characteristics, 46; its extreme liability to decomposition and volcanic nature, 47; hypotheses of various naturalists on the subject, 47-49; odour given out by it when rubbed, 53; M. le

FOSSILS.

Coq's hypothesis relative to the domitic puys, 70 n.
Dordogne, 4, 5, 114, 126, 127, 132, 134 n., 135, 137, 139, 140, 146, 149.
Dordogne, Lot, and Aveyron, character of the lias underlying the oolite in the departments of, 5.
Dore, Mont. *See* Mont Dore.
Dore, the, 112, 116.
Dore and Dogne rivers, point of union of the, 116, 124, 128.
Douc, Montagne de, 168, 171, 174.
Drachenfels, 130.
Drôme, 1.
Durbise, la, 128.
Durtol, 64, 104.

Egaules, 67.
Egravats, Ravin de, 131 n.
Eiffel, 78, 81.
Emblavès, 26, 178, 179.
Enfer, Puy and Vallée de l', 94, 126.
Entraigues, 194.
Enval, 89.
Eraignes, Puy d', 143.
Erieux, 162.
Escobar, M., 116 n.
Escoutay, 165.
Espinasse, la Narse d', 94.
Etang de Fung, 59.
Etang, l', 83.
Eternity. *See* Time.
Etna. *See* Ætna.
Euganean Hills, 130.
Expailly, 171, 181, 182 n.
Eysenac, Montagne d', 180.

Falgoux, 147.
Faujas de St. Fond, M., 31, 32, 165, 187, 192, 245.
Fauna of Central France. *See* Organic remains.
Fay, 179.
Fay-le-Froid, 167.
Ferrand, Puy, 124, 125, 133.
Ferrara, M., 116 n.
Filhou, Puy de, 61.
Fontanat, 86.
Fontaulier, the, 185, 187.
Font de l'Arbre, 86.
Fontfrede, 89, 90, 91.
Fontmore, 64.
Forèz, the, 1, 2, 7, 28, 210. *See* Auvergne and Forèz.
Fortresses, ruined, on the peaks of the Auvergne, and ultimate fate of their occupants, 102 n.
Fossils and organic remains of the Haute Loire, 229, 230. *See* Organic remains.

Fournet, M., 35 n.
Fraisse, Puy de la, 66, 67, 70.
France, Central, division line of, 1; geological features of the country on either side of this line, 1-5; theory of the French geologists relative to the production of its volcanic rocks, 200; table of heights of the volcanic districts, 232.
Freshwater lakes of the tertiary period, proofs of the former existence of, 2, 6; probable cause of their drainage, 6; their boundaries easily recognisable, *ib.*; lakes of La Caissière and d'Aidat, 92; peculiar features and probable origin of Lakes Pavin and Mont Sineire, 143, 144; question of the original level of the lake-basins of Central France, 210-213. *See* Cantal, Haute Loire, Limagne d'Auvergne, Menat, Montbrison.

Gannat, 10, 12.
Gard, 3, 4.
Garges, 4.
Geology of Central France in the regions of the Auvergne, Forèz, Gevaudan, Vivarais, and the Valley of the Rhone, 1, 2; freshwater deposits of the tertiary period, 2; crystallizations in the granite, *ib.*; metals and their localities, 2, 3; purity of the kaolin of Limoges, 3; absence of the Cambrian, Silurian, and Devonian series of strata, 3; coal-measures and their localities, 3, 4; extent of limestone strata of the lias and oolite group, 4; aspect of the "causses" in the region of the Cevennes, 4, 5; lesson taught by geology relative to the immensity of the world's epochs, 208, 209 and *n.*; Sir Charles Lyell's summary of the geological characteristics of Central France, 213, 214. *See* Organic remains, Volcanic formations, Volcanic remains, Volcanic rocks.
Gerbier des Joncs, 157.
Gerbison, 179.
Gergovia, 12, 15; volcanic constituents of the mountain of, 17, 18, 200, 203; geological and antiquarian interest attaching to it, 106 and *n.* †; difference between its basalt and that of La Serre, 204. *See* Girou.
Gevaudan and Vivarais, surface elevation attained in the region of the, 1.
Gimeaux, 23.
Gironde, 1.
Girou, Puy, 12; this puy and Puys de Jussat and Gergovia originally a single plateau, 106; character of the stratification of the region, 107; line of contact between the basalt and the limestone well defined, 108; its altitude compared with that of adjacent puys, 109; its height, 233.
Godivel, La, 144.
Goul, the, 146.
Goules, Puy des, 67, 69, 70; its height, 233.
Gour de Tazana. *See* Tazana.
Gouette, Puy de la, 71, 72, 73, 75.
Grange, Pan de la, 125.
Grange, Puy de la, 132.
Granite rocks of Central France, varying character of the, 2.
Graveneire, Puy, 44, 83, 99*n.*, 100; its puzzolana in much request, 84; non-existence of a crater, *ib.* and *n.*; character of its lava rock, 85; its gaseous springs, *ib.*; industrious cultivation of the district, 86; its height, 233.
Gravouse, Puy de la, 91, 92.
Gresinier, 64, 105.
Griou, Puy, 148.
Gromanaux, Puy de, 87.
Gros, Puy, 136, 137; its height, 232.
Grotta del Cane, French springs analogous to the, 85, 193.
Guery, Lake, 137.
Guettard, M., an early observer of the volcanic phenomena of Central France, 30; small credit given to his memoir thereon, 31.
Guiolle, La, 39, 150.

Hamilton, Sir W., on the lava-current of Ætna, 117 *n.*
Hamilton and Strickland, Messrs., on the volcanos of Asia Minor, 210 *n.*
Hautechaux, Puy de, 133.
Haute Loire, freshwater formation of the basin of the, 25; extent and depth of the superimposed volcanic rocks, *ib.*; limits of the original basin, outlets of the Loire, 26; constituents of the lower series of lacustrine beds, 26, 27; extent and variety of their organic remains, 27; possible cause of the drainage of the lake, *ib.*; points of resemblance between the rocks of this district and the peperino of the Auvergne lake basin, 28; probable cause of the accumulation of the water into a lake, 179.
  VOLCANIC REGION of the HAUTE LOIRE and ARDÈCHE (ci-devant provinces of the *Velay* and *Vivarais*), 154; natural boundary of this district and the Cantal, 155; its calcareous formation entirely cased in granitic rock, 198; region of Mont Mezen and its dependencies, *see* Mont Mezen. Volcanic

HAUTES ALPES.
phenomena of the Coiron, *see* Coiron. Region of the Velay and Vivarais, *see* Velay. *See* also 200, 206, 210.
Hautes Alpes, 1.
Herault, 4.
Herculaneum and Pompeia, 30.
Human bones, discovery of, in the breccia of the Montagne de Denise, 181, 182; results of a scientific discussion of the subject, 183; inferences deducible from the discovery, *ib.*
Humboldt, 51, 122, *n.*

Iceland, resemblance of the Mont Baula in, to the Puy de Dôme, 50 *n.;* enormous extent of the lava-current of a volcano in, 117 *n.;* bursting of trachytic domes and currents through basaltic beds in, 201 *n.*‡
Incrusting springs, natural bridges formed or forming by, 22; localities of similar springs, 23; different character of their sources, *ib.*
Infau, Puy de l', 94.
Ischia, 130.
Isère, 1.
Issarles, Lake d', 185.
Issingeaux, 173.
Issoire, 2, 211.
Italy, 30 *n.*; extinct volcanos of, 119 *n.* †; natural parallel to the Roman roads in, 189.

Jaligny, 12.
Jaujac, la Coupe de, 186, 191, 193; notion countenanced by its coal formations, 190; its chesnut-trees, gaseous spring, and beautiful basaltic range, *ib.*
Joncs, le Gerbier des, 157.
Jourdanne, the, 25, 148, 149.
Joyeuse (Ardèche), 4.
Jughat, Mont, 12.
Julius Cæsar, site of an unsuccessful siege by the legions of, 106.
Julliat, 91.
Jume, Puy de, 68, 77; its height, 233.
Jumes, 100. *See* Châteaugay.
Jussac, 25.
Jussat, Puy de, 12, 106. *See* Girou.

Kaolin, supply of, for the factories of Sèvres and Paris, whence procured, 3.

Lacoste, M., on the volcanos of Auvergne, 34.
Lacustrine formations. *See* Freshwater lakes.
Laduegne, 165.
Laizer, M. de, 35.

LIMAGNE.
Lakes and lake basins. *See* Freshwater lakes.
Lamoreno, Puy, 89.
Lancerote (a Canary Isle), references to volcanic eruptions in, 95 *n.**, 180 *n.*
Langogne, 184.
Languedoc, 186, 197.
Lantegy, Puy de, 71.
Lantriac, 173.
Laqui, 77.
Lardeyrolle, 159.
La-roche-lambert, 169.
Laschamp, Puy de, not volcanic in appearance when seen from a distance, 89; marked effects of fire observable on this hill, 90; its height, 232.
Lavas, laws determining the propulsion of, 49-51; rock of prismatic lava near Pont Gibaud, 61; great extent of various lava-currents, 117 *notes*; typical character of the various lava, 130 *n.*
Laveille, 135.
Lavoulte (Ardèche), 4.
Layssac (Aveyron), 4.
Le Coq, M., 35, 70 *n.*, 87 *n.*
Le Crest, village of, 102, 104.
Le Grand d'Aussy's 'Voyage en Auvergne,' 33.
Leironne, Puy de, 71.
Lempde, 4, 154.
Le Puy. *See* Puy, Le.
Le Vernets, 2.
Limandre, Lake de, 185.
LIMAGNE D'AUVERGNE, 2, 24, 25, 27, 38, 39, 64, 82, 112, 134, 198, 200, 202, 205, 210, 211, 212, 213; extent and characteristics of its lacustrine formation, 7, 8; principal divisions of the lacustrine series, the sandstones and conglomerates, and M. Brongniart's opinion thereon, 8; the author's deductions on the same subject, *ib.*; identity of the marls and sandstone with the secondary *new red* sandstone and marl of England, 9; character of the green and white foliated marls, 9, 10; remarkable extent covered by the indusiæ of the Phryganea (or caddis-fly), 11, 12; localities where they are best developed, 12; probable cause of their aggregation, *ib.*; components of the calcareous marls, 13; M. Pomel's deductions relative thereto, 13, 14; coincidence of Sir Charles Lyell's opinion therewith, 14; calcareo-volcanic strata of the district, 15-17; calcareous peperino, 18-21; calcariferous and incrusting springs and their localities, 21-23.
LIMAGNE and MONTS DOME, VOLCANIC REGION of the, its extent and elevation,

LIME.

40; its components and external characteristics, 41, 42; aspect of the lava-fields, 43; epochs of the formation of the *chain* of *puys*, see Puys, chain of.

Lime, limestone, carbonate of lime, producing causes of, in the regions described in this work, 198, 199.

Limoges, 3, 67.

Lipari Isles, 48.

Loire, the, 1, 2, 3, 4, 7, 25, 26, 27, 28, 157, 167, 171, 175, 176, 177, 178, 179, 207, 211.

Loire, Upper. *See* Haute Loire.

Lot, 3, 5, 146.

Louchadière, Puy de, 56, 58; its majestic proportions, 78; appellation suggested by its crater, 79; its height, 233.

Loueire, Puy de, 136.

Louis XIV., 5, 102 *n.*

L'Oulette, 178.

Lozère, 3, 4, 154 *n.*; height of Mont Lozère, 232.

Lyell, Sir Charles, references to the writings of, 9, 12, 14, 15 *n.*, 27, 35, 96 *n.*, 116 *n.*, 122 *n.*, 135 *n.*, 199, 217, 223, 225, 226; his summary of the geological characteristics of Central France, 213, 214.

Lyonnais, 28.

Lyons, 150.

Macalouba, 122.

Madeira, 116 *n.*

Malesherbes, M., one of the earliest observers of the volcanic phenomena of Central France, 30, 31.

Man, coexistence of, in France, with races of extinct mammalia, 183.

Manson, Puy de, 233.

Manzat, 80.

Marais, Dent du, 135.

Marcilly, 28.

Margeride, La, 149, 154, 184.

Mari, Puy, 149; its height, 232.

Marine deposits, none later than the Jurassic system in the regions described in this work, 197.

Marmant (or Marmont), Puy, 18, 20.

Marone, the, 146.

Mars, the, 146.

Marsat, 83.

Massiac (Cantal), 3.

Mauriac, 4.

Mayence, 14.

Mayet d'Ecole, 12.

Mazayes, 58.

Mazayes, lake of, 59.

Mercœur, Puy de, 90, 158, 173.

Meerfeld, 81.

MONT DORE.

Menat, tripoli basin of, 28; result of excavations in its sedimentary beds, 29; origin of its tripoli, *ib.*

Meye, Puy de la, or Puy Noir, 90, 91, 103.

Mezen, 38, 49, 125 *n.*, 175, 177, 180, 194. *See* Mont Mezen.

Mianne, 161 *n.*

Miaune, 179.

Millefleur, Puy de, 13.

Modena, 122.

Monges, the, 59.

Monistrol, 184.

Mont, Montagne de, 180.

Montaigu, 28.

Montaudou (or Montaudoux), Puy de, 84, 99; its composition, 99 *n.*

Montbrison, 2; extent and character of its basin, 28; its calcareous formation, 198; original limits of its lake basin, 210.

Montbrul, les Balmes de, an accidental excavation, 165.

Mont Chagny, 12.

Montchal, Puy de, 92, 143.

Montchar, Puy, 89; its height, 232.

Montchie, Puy de, 42, 88.

Mont Crousteix, 171.

Mont Dore, 33, 34, 38, 47, 48, 49, 50, 67, 70 *n.*, 78, 82, 89, 95, 110, 146, 147, 148, 149, 152, 155, 156, 160, 167, 199, 201, 209, 211. REGION OF MONT DORE. I. GENERAL OUTLINE: its figure, 114; its ancient appellation, *ib. n.*; points of similarity between it and other insulated volcanic mountains, 115; circumstances which would reduce Ætna to the condition of Mont Dore, 116 *n*; relative positions of its basaltic and trachytic products, 116, 117; their occasional approaches to similarity of appearance, 118; extent, condition, and character of its conglomerates, 118–121; facts tending to prove the co-agency of water in the formation of these conglomerates, 121, 122; causes of deluges during volcanic eruptions, 122; effects naturally producible thereby, 123, 124; large annual fall and long continuance of snow on Mont Dore, 123 *n.* II. ITS STRUCTURE: heights of Pic de Sancy and Puy Ferrand, and prospect therefrom, 124; nature of the rock composing the mountain platforms, 125; sulphur and alum deposits beneath the Cascade of the Dore, *ib.*; rich pasturages of the high-lands, marked contrast between the valleys and the heights, 125 *n.*; les vallées de l'Enfer and de la Cour, and their stratification, 126; locality of the principal vent (central crater),

MONT DORE.

whence issued the trachyte formations of the mont, 127; le Cliergue and its beds of building-stone, 128; height, slope, directions, and succession of the bed watered by the Cascade du Mont Dore, 128-131; composition of the Plateaux de Rigolet, Bozat, and Charlanne, and the Puy de la Grange, 132, 133; highest points and average thickness of the granitic substratum of Mont Dore, 134 *n.*; valleys of Chambon and Besse, and bone-beds of Mont Perrier, 134, 135; characteristics of the *Northern Flank*, 135 – 139; *Southern Flank* — greater abundance of basalt, 139; indefiniteness of the limits and scantiness of vegetation of this district, 139, 140. III. RECENT VOLCANIC ERUPTIONS: site and composition of the Puy de Tartaret, 140, 141; decisive proofs here afforded of the igneous origin of basalt, 142 and *n.*; characteristics of the basalt of Tartaret, 143; peculiar character and probable origin of the Lakes Pavin and Mont Sineire, 143, 144.

Mont Dore, Cascade of, 125, 129, 131 *n.*
Mont Dore les Bains, 23, 118, 129.
Montelimart, 30.
Montenard, Puy de, 95, 140.
Montgy, Puy de, 93; its height, 233.
Monti Cimini, 48.
Montilhe, Roc de la, 138.
Montillet, Puy de, 93.
Monti Rossi, 65.
Montjoly, 84; its gaseous springs, 85.
Mont Jughat, Puy de, 12, 94.
Montlosier, M. de, 35, 37, 48, 89; his agricultural labours, cause of his burial in unconsecrated ground, 238.
MONT MEZEN and its dependencies, characteristics of the volcanic region of, 155; causes of the diversity of aspect between this and other regions, 155, 156; predominance of *clinkstone* in its composition, 156; extent covered by the clinkstone, 157; strata on which the clinkstone rests, 158, 159; nature and varieties of the clinkstone, 160, 161; debateable origin of the clinkstone hills, 161 *n.*; extent and flow of the basaltic currents of the district, 167; opinion of the author relative to their origin, *ib.*; facts whereon such opinion is based, 167-172; basaltic dykes and their formation, 173-175; its height, 232. *See* also 199, 209.
Monton, Puy de, 120, 123.
Montpensier, Butte de, 13.
Mont Perrier, 14, 135.
Montpézat, 185, 189, 192.
Montpézat, la Gravenne de, 186; extent and

ORGANIC.

depth of the crater of, 187; beauty of its basaltic columnar ranges, *ib.*
Montplaux, 27.
Montredon, castle of, 102, 104, 181.
Montrognon, probably a remnant of a current of Gergovia, 108; its altitude compared with that of adjacent puys, 109.
Mont Serae, 178.
Mont Sineire, 143.
MONTS DÔME, 201, 205, 210. *See* Puys, chain of.
Mossier, M., opinion of, on the formation of the puys, 48.
Moulins, 4, 7, 197, 211.
Mozun, 112, 113.
Mur de Barrez, 25.
Murat, 25, 148, 149, 150, 151.
Murat le Quayre, 134 *n.*, 137.
Murchison, Sir Roderick Impey, 35.
Murol, 142.

Nadailhat, 102.
Naples, 78, 120; Phlegræan fields of, 119.
Napoleon, 154 *n.*
Nechers, 120, 135, 141.
Neidermennig, 78.
Neptunians and Vulcanists, source of the contest between the, 101 *n.* *See* also 142 and *n.*
Néris, 23.
Neufort, Puy de, 87.
Nevers, 197.
Neyrac, 193.
Niaigles, 91.
Nièvre, 4.
Nohanent, 64.
Noir, Puy (or Puy de la Meye), 90, 91, 103, 203.
Nugère, Puy de la, 77, 80, 83, 118, 128, 201; its quarries of building-stone, 78.

Olby, 89.
Olivine not an invariable characteristic of basalt, 130 *n.*; beauty of the olivine of Burzet, 188.
Olloix, Puy d', 110.
Opme, village of, 235.
Orcines, 63; its height, 233.
Organic remains of Central France: preliminary remarks, 217, 218; fauna of the lacustrine strata, 218-223; pliocene fauna of the volcanic era, 223, 224; locality to which this latter fauna is almost wholly confined, 225; fish of the Basin of Menat (qu. post pliocene?), *ib.*; fauna of ancient alluvia (post pliocene), 226, 227; periods to which the alluvial fauna is referred by M. Pomel, 228; M. Aymard's classification, 229, 230;

observations on M. Aymard's catalogue, 230, 231.
Orgues, the, 181.
Ours, Mont d', characteristics of, 241.

Paille, Croix de la, 181.
" Palais du Roi," les, 164.
Palisse, la, 12.
Palma, 115, 116 n.
Pandreaux, les, 173.
Panet, 25.
Pardines, 14, 120, 135.
Pariou, Puy de, presumably the product of one of the last eruptions, 61; dimensions of its crater, 63; interesting character of its lava flow, 63, 64; extreme ruggedness of its " cheire," 65 ; accumulation of puzzolana probably due to its eruptions, 66 ; its height, 232. *See* also 67, 68, 86, 105.
Paris, 3, 150, 212.
Pasredon, Puy de, 109.
Passis, M., 35 n.
Paulhaguet, 155, 176.
Pauniat, Puy de, 79.
" Pavés de Géans," 188.
Pavin, lake, 143 ; its height, 232.
Pennant, extent of an Icelandic lava-current described by, 117 n.
Peperino. *See* Calcareous peperino.
Perignat, 108.
Pessade, Puy de, 133.
Petrifying springs. *See* Incrusting springs.
Peylenc, Rochers de, 169.
Pèze, le Cros de, 137.
Pezenas, 196.
Phlegræan fields of France, 157; of Naples, 119.
Phonolitic mountains, tendency of, to waste into detached conical masses, 159.
Phryganea (or caddis-fly), remarkable extent of territory covered by the indusiæ of the, 11, 12; localities where they are best developed, 12.
Pierre Neyre, 178, 179.
Pierre-sur-Haute (Forèz), 232.
Piquette, Puy de la, 20.
Pissis, M., idea of, relative to the superposition of basaltic currents, 201 ; his opinion on the elevation of the freshwater formation of the Limagne, 211.
Planèze, la, 149.
Plomb du Cantal, 148 ; its height, 232.
Poix, Puy de la, 18, 21.
Polignac, Rocher de, 172, 175.
Polminhac, 25, 151.
Polognat, 120, 138.
Pomel, M., 13, 14, 27, 35 n, 135, 217; his fauna of the lacustrine strata of Central France, 218, 223 ; pliocene fauna of the volcanic æra, 223-225; fauna of ancient alluvia, 226-228.
Pont Barraud, 12,
Pont de Baume, 192.
Pont des Eaux, 89.
Pout du Château, 15, 18, 20, 40, 112, 200.
Pont Gibaud, 3, 23, 59, 60, 79, 96 ; sketch of rock of prismatic lava near, 57.
Ponza, 48.
Poulet, Puy de, 133.
Pourcharet, Puy de, 93.
Pourseille, 187.
Pradelle, 155, 176, 185.
Prades (Ardèche), 4.
Prevost, M., 35 n.
Prezailles, 176.
Privas, 162.
Prudelle, Plateau de, 104; decomposed character of its scoriæ, 105; difference between the two species of basalt of which its current consists, 105, 106 ; its height, 233.
Puy, Le, 20, 25, 26, 158, 166, 168, 171, 172, 173, 178, 180, 181, 182, 183, 185, 199, 207, 210.
Puy, Petit, or Petit Puy de Dôme, and its " Hen's Nest," 54, 55 ; its height, 232.
PUYS, CHAIN OF, OR MONTS DÔME, situation of the, 39 ; epochs of their formation, 44, 45 ; hypotheses of various naturalists relative to such formation, 47–49 ; laws determining the propulsion of lavas, 49-51 ; Von Buch's theory, its strong and weak points, 51 ; puys north of Puy de Dôme, 54-82 ; eastern line of puys, 82-86 ; puys facing Puy de Dôme, 86, 87 ; puys south of Puy de Dôme, 87-95.
" Pyramids of Botzen " (Tyrol), cause of the preservation of the, 173 n.
Pyrenees, the, 186.

Querilh, Cascade de la, 131.
Queuille, La, 136.

Rambon, la Fontaine de, 23, 198.
Ramond, M., 7, 34, 35, 48, 51, 67, 99, 110, 114, 119, 120, 142, 213, 232.
Rantières, 144.
Raulin, M., 28, 35 n., 111, 152, 201 n., 211, 212 n.
Recupero's notice of a lava-current of Ætna, 117 n.
Rhone, the, 1, 3, 31, 164, 165, 166, 167, 193.
Richelieu, Cardinal, 102 n.
Rigolet, Plateau de, 128, 131, 132.
Riom, 12, 28.

Rioupezzouliou, the, 171.
Rive de Gier (Loire), 3, 4.
Roanne, 2, 7.
Robert, M., Icelandic mountain described by, 50 n.
Roche, La, 185.
Roche en Regnier, 157.
Rochefort, 136, 138.
Rochemaure, 31, 164.
Roche Noire, Puy de la, 112.
Roche Rouge, 173, 174, 175.
Rocher Corneille, 172, 175, 178.
Rocher de Polignac, 172.
Rochers de Peylenc, 169.
Roches, les, 84.
Rodde, Puy de la, 94.
Romagnat, 12.
Romans, remains of the, 85, 98, 106 and n. †.
Rome, 31, 119.
Ronzon, 27.
Rouge, Puy, a connecting link between recent and early eruptions, 96; site of its orifice, 97; inferences deducible from the deep excavation formed by the river, *ib.*; its ancient and modern lead-mines, 97, 98; its "pièces séparées grenues" not the product of decomposition, 98.
Roulade, Puy de la, remarkable feature of the, 100; epoch of its formation, *ib.*
Roux, M., 232, 246.
Royat, bed of basalt in the valley of, 84; its medicinal springs, 85.
Rozet, M., 35 n., 200.
Rozières, 173.
Rue, the, 146.
Ruelle, M., 35 n., 148 n.

St. Agrève, 167.
St. Alyre, 23, 198.
Ste. Anne, Montagne de, 180.
St. Arcize, 152.
St. Arçon, 184.
St. Bonnet, 26, 138.
St. Eloy (Puy de Dôme), 4.
St. Etienne, 3, 4, 178.
St. Floret, 23.
St. Flour, 149.
St. Front, Lake de, 185.
St. Geneis, 152.
St. Genest de Champanelle, circumstance most remarkable in the basalt of, 109.
St. Geneste, 83.
St. Geneys, 184.
St. George d'Aurat, 185.
St. Gerard-le-Puy, 12.
St. Hippolyte, Puy, 113.
St. Ilpize, 184.
St. Jean, 165.

St. Jean le Noir, 243.
St. Michel, 172, 175.
St. Nectaire, 23, 198.
St. Paul de Salers, 147.
St. Pierre Eynac, 27, 158.
St. Privat, 185.
St. Romain Turluron, 113.
St. Sandoux, Puy de, 110, 111; its height, 233.
St. Saturnin, 92, 110.
St. Sauve, 137.
St. Vincent, 77.
St. Yrieix, 3.
Salers, Montagne de, 149.
Salomon, Puy de, 88, 89.
Sanadoire, 136, 137.
Sancy, Pic de, 114, 123, 124, 125; its height, 232.
Sansac (Aveyron), 4.
Saône et Loire, 4.
Sarcouï, Puys de (Great and Little), 67, 68, 69; and Puy des Goules—mineralogical features and wintry dangers of the latter, 67; comparison suggested by the shape of the Grand Sarcouï, 68; advantages derivable from the study of this group, 69; height of the Grand Sarcouï, 233.
Sault, Puys de, Petit et Grand, 87.
Saussure's opinion on the Monts Dôme, 47.
Sayat, 77.
Scie, La, valley of, 131, 132.
Scotland, shell-marl deposits of, and calcareous freshwater deposits of Central France, 199.
Scrope's, G. P., 'Considerations on Volcanos,' references to, 49 n., 51, 65 n., 117 n., 122 n., 123 n., 188 n., 192 n.
Seine, the, 1.
Seinzelle, Montagne de, 180.
Serre, Puy de la, 12.
Serre, la, Plateau of, circumstance to which its chief interest is due, 101; its basaltic bed a battle-ground for the Neptunians and Vulcanists, *ib.* and n.; origin of its current, 102; parallelism of its older and newer currents, 103; probable results of already existing excavations, *ib.*; scientific use of this hill as a type of the formation of basaltic plateaux, 104, 203; difference between its basalt and that of Gergovia, 204; its height, 233.
Servissas, 174.
Sèvres, source of supply of kaolin for the china factories of, 3.
Sidonius Apollinaris, locality of a residence of, 92 n.*
Siebengebirge, 29.
Sioule, the, 7, 29, 40, 56, 57, 58, 59, 87,

89, 96, 138; sketch of rock of prismatic lava on its banks, 57.
Solas, Puy de las (or Puy de la Gravouse), 91, 92; its height, 232. *See* Vache, Puy de la.
Souillols, la Gravenne de, 186; peculiarity of its twofold columnar range, 191; probable cause of their difference of structure, 191, 192; its gaseous spring, 193.
Spallanzani, dimensions of lava-currents mentioned by, 117 *n*.
Springs, calcariferous. *See* Calcariferous springs.
Springs of carbonic gas at Mont Joly and Royat, 85; at Jaujac, 190; at Souillols, 193.
Springs, incrusting. *See* Incrusting springs.
Suchet, Great and Little, 55, 66, 67, 332.
Sumène, the, 170, 178, 179.
Sury le Comtal, 28.
Switzerland, 32, 212.

Tache, Puy de la, 133.
Tallande, 92.
Tarare, 3, 28.
Tartaret, Puy de, 140, 141, 142, 143.
Taupe, Puy de la, 94.
Tazana, le Gour de, peculiarity of its crater, 81; its probable origin, 82. *See* also 143, 185.
Teneriffe, Peak of, 38, 115; condition of its flanks, 116 *n*.; bursting of trachytic domes and currents in, 201 *n*. ‡.
Ternant, 66.
Theix, Valley of, 90.
Thiezac, 152.
Thiolet, 79, 80.
Thueyts, volcano of, 186; its character, 189; its majestic colonnade of basalt, *ib*.
Time considered in reference to geological epochs, 208, 209; ideas suggested by the terms eternal, eternity, for ever, 209 *n*.
Toulon, 196.
Touraine, 14, 197.
Trachytes of Mont Dore and Cantal, 49, 50; trachytic domes of Iceland and Hungary, 50 *n*.; trachytic volcanos of America, 122; possibility of establishing a distinguishing feature between basalt and trachyte, 130 *n*.; question of the alternate superposition of the two, 131 *n*. *See* Domite, Mont Dore.
Travertin, natural bridge of, at Clermont, 22.
Tripoli basin of Menat, 28; probable origin of the tripoli, 29.
Tuilière, La, 136, 137.
Tyrol. *See* Pyramids of Botzen.

United States, kaolin exported from France to the, 3.
Usclades, 185.

Vache, Puy de la, and Puy de las Solas, created by showers of scoriæ previous to any emission of lava, 91; peculiar condition of the clay in contact with the basalt, 92 and *n*. †; scoriæ of Puy de la Vache, 93.
Vals, 194.
Vandeix, Valley of, 132.
Vauquelin's experiments on domite, 53.
Vayre, 20.
Vegetable remains. *See* Organic remains.
Velay, 2, 32, 34, 38, 60, 100, 155, 166; vestiges of deposits of freshwater lakes in the, 2; chain of recent volcanos in the Velay and Vivarais (*Haute Loire* and *Ardèche*), 175; extreme numerousness of the scoriæ cones, 176; effects produced by the lava-currents upon the course of the Loire, 177, 178; M. Aymard's view relative to the subject, 178 *n*.; phenomena connected with gorge of the Sumène, 178, 179; parallelism of the volcanic eruptions of Auvergne and the Velay to those of Lancerote, 180 *n*.; the Montagne de Denise and the deposits of human bones in its breccia, 181, 182; results of a scientific discussion of the subject, 183; mural ranges of columnar basalt at St. Ilpize, &c., 184; sites of lakes on this range, 185.
Velay, Puy en, 154.
Vernets, Le, 2.
Verthaison, 18, 19.
Vesuvius, 30, 31, 44, 127, 134 *n*.
Vic en Carladez, 23, 25, 151.
Vicentin, 19 *n*.
Vichatel, Puy de, 92, 93.
Vichy, 23.
Vieille Brioude, 154 *n*., 184.
Villar, 63, 64, 104, 105.
Villefort (Ardèche), 3.
Villefranche (Allier), 4.
Villejacques, 139.
Villeneuve, 163.
Villeneuve de Berg, 164.
Viol, Puy de la, 87.
Violan, Puy (Cantal), 232.
Vivarais, 1, 31, 32, 34, 38, 39, 58, 100, 155, 163 *n*., 178 *n*., 184.
Vivarais, Bas, 166, 178; interest attaching to its recent volcanic remains, 186; magnificence of its slopes and valleys, *ib*. and *n*.; its six volcanic cones, *see* Ayzac, Burzet, Jaujac, Montpézat, Souillols, Thueyts; probable point of union of its

five lava-currents, 192; destructive process continually operating on the basaltic and granitic beds of these valleys, 192, 193; presumed cause of the preservation of the volcanic remains of this region, 194; characteristics of the basaltic dyke near Aubenas, 195, 196; proofs deducible from the lavas of this region, 205.

Vivarais, Haut, 26, 157, 162, 166, 185, 193.

Viviers, 243.

Volant, the, 193, 194.

Volcanic "cone of eruption," explanation and description of a, 41 *n*. †.

Volcanic formations on the elevated granitic platform of Central France—theories of Dr. Daubeny and M. Montlosier, 37; the author's reasons for dissenting therefrom, *ib.*; groups into which the entire district is geographically divisible, 38, 39; volcanic formation of the Velay and Vivarais, *see* Haute Loire, Coiron, Mont Mezen, Velay.

Volcanic regions: I. *See* Limagne d'Auvergne, and Puys, chain of. II. *See* Mont Dore. III. *See* Cantal. IV. *See* Haute Loire.

Volcanic remains of the interior of France, 30; ignorance regarding them previous to 1750, *ib.*; Researches of MM. Guettard and Malesherbes, 30, 31; discouraging reception of M. Guettard's memoir on the subject, 31; his Clermont opponent, *ib.*; results of the labours of MM. Desmarest and St. Fond, 31, 32; accuracy of Desmarest's maps, points wherein both naturalists were in error, 32; Dolomieu and his igneous theory, *ib.*; character of Le Grand d'Aussy's 'Voyage en Auvergne,' 33; M. de Montlosier the first to establish the true nature of these remains, *ib.*; MM. de Buch's, Lacoste's, and Ramond's writings on the subject, 33, 34; M. d'Aubuisson's observations and their result, 34; value of Baron Ramond's measurements and observations, 34, 35; later writers hereon—Dr. Daubeny, MM. le Coq, &c. &c., and Sir C. Lyell and Sir R. Murchison, 35 and *n*.

Volcanic rocks, no traces of, in the lower terms of the Limagne freshwater formation, 14; characteristics of the calcareovolcanic strata of that district, 15-17; attempts of French geologists towards a periodic classification of the volcanic rocks of Central France, 200; facts and observations demonstrative of the intenability of such a classification, 201-207; immensity of the periods required for the production of the various effects observable in geological surveys, 208, 209 and *n*.; probable periods of the most remarkable of the volcanic products of Central France, 209, 210; analogy between them and the volcanic rocks of Asia Minor, 210 *n*.

Volcanic rocks, the product of earlier eruptions, elucidatory observations on, 95, 96. *See* Berzé, Charade, Chanturgue, Châteaugay, Coran, Cornon, Gergovia, Girou, Prudelle, Roulade (Puy de la), Rouge (Puy), St. Genest de Champanelle, Serre (la, Plateau of).

Volcanos of Central France, notices hitherto published of the. *See* Volcanic remains.

Volvic, 78, 79, 83, 118, 128.

Vulcanists and Neptunians, source of the contest between, 101 *n*. *See* also 142 and *n*.

Weiss, Dr., 137.

Werner, his theory and his school, 142 *n*.

Yolet, 151.

29 AP 58

THE END.

LONDON: PRINTED BY W. CLOWES AND SONS, DUKE STREET, STAMFORD STREET, AND CHARING CROSS.

ALBEMARLE STREET,
October, 1857.

MR. MURRAY'S
List of Works in the Press.

A MEMOIR OF THE REMARKABLE EVENTS
WHICH ATTENDED THE ACCESSION TO THE THRONE OF THE LATE
EMPEROR NICHOLAS I. OF RUSSIA.

Drawn up under his own inspection, by BARON M. KORFF, Secretary of State, and now published by Special Imperial Command.

8vo. 10s. 6d. (*Ready.*)

Published simultaneously in RUSSIA, ENGLAND, FRANCE, and GERMANY.

INDIA.
LETTERS, DESPATCHES, AND OTHER PAPERS.
BY FIELD MARSHAL THE DUKE OF WELLINGTON.
NOT HITHERTO PUBLISHED.

Edited by the PRESENT DUKE.

Uniform with the First Edition of the WELLINGTON DESPATCHES, edited by COLONEL GURWOOD.

Two Vols. 8vo.

These volumes consist of Documents relating to INDIA, discovered since the death of the DUKE OF WELLINGTON, and of a few which were printed in the Second Edition of "GURWOOD'S DESPATCHES," but not included in the First Edition.

MISSIONARY TRAVELS AND RESEARCHES IN SOUTH AFRICA;

Including a Sketch of Sixteen Years' Residence in the Interior of Africa, and a Journey from the Cape of Good Hope to Loanda on the West Coast; thence across the Continent, down the River Zambesi, to the Eastern Ocean.

BY DAVID LIVINGSTONE, LL.D., M.D., D.C.L.

With Portrait, Maps by ARROWSMITH, and numerous Illustrations. 8vo. (*On Nov. 10th.*)

ON THE RIGHT USE OF THE EARLY FATHERS:

A Course of Lectures delivered in the University of Cambridge.

By Rev. J. J. BLUNT, B.D.,
Late Lady Margaret's Professor of Divinity.

8vo. Uniform with Blunt's "History of the Christian Church."

THE GEOLOGY AND EXTINCT VOLCANOS OF CENTRAL FRANCE.

By G. POULETT SCROPE, M.P., F.R.S., F.G.S., &c.

Second Edition, enlarged and improved, with additions and corrections. With Maps, Views and Panoramic Sketches. Medium 8vo.

The Author, having recently revisited Central France, has been enabled to revise his work, and to add the results of his researches and those of other Geologists up to the present time.

HISTORY OF HERODOTUS.

A new English Version, from the Text of GAISFORD, Edited with Copious Notes and Appendices, illustrating the History and Geography of Herodotus, from the most recent sources of information, embodying the Chief Results, Historical and Ethnographical, which have been obtained in the Progress of Cuneiform and Hieroglyphical Discovery.

By REV. GEORGE RAWLINSON, M.A., Exeter College, Oxford.

ASSISTED BY

COL. SIR HENRY RAWLINSON, K.C.B., and SIR J. G. WILKINSON, F.R.S.

With Maps and Illustrations. 4 Vols. 8vo.

ROUND THE ROCK.

Excursions into Andalusia and the Coast of Africa from the Rock of Gibraltar, with some Notices of the Spanish Pyrenees. In a Series of Letters.

By JOHN ADOLPHUS, M.A., Barrister-at-Law.

Post 8vo.

GALLERIES AND CABINETS OF ART IN ENGLAND.

Being an Account of more than Forty Collections, visited in 1854 and 1856, and now for the First time Described.

By DR. WAAGEN, Director of the Royal Gallery of Berlin.

With Index. 8vo.

Forming a Supplemental Volume to his "Treasures of Art in Great Britain."

"Dr. Waagen's admirable volumes."—*Christian Remembrancer, Oct.,* 1857.

A NEW HISTORY OF MODERN EUROPE,

From the taking of Constantinople by the Turks to the Close of the War in the Crimea.

By THOS. H. DYER.

Author of "The Life of Calvin," the article "Rome" in Smith's Dict. of Classical Geography, &c. &c.

4 vols., 8vo.

Much light has been thrown during the last half century on the modern history of Europe; yet Dr. Russell's work upon the subject, although quite antiquated and frequently deficient in correctness, remains the only one to which the English reader can have recourse. The present work, founded on the best and most recent historical publications, both English and foreign, has been undertaken in order to supply an obvious want.

THE SEPOY REVOLT;

ITS CAUSES AND ITS CONSEQUENCES.

BY HENRY MEAD.

8vo.

MANUAL OF THE NATURAL HISTORY OF FOSSIL EXTINCT ANIMALIA.

The substance of a course of popular Lectures delivered in the Museum of Economic Geology, in the Spring of 1857.

By RICHARD OWEN, F.R.S.

With Illustrations. 8vo.

Uniform with Lyell's "Manual of Elementary Geology."

SICILY;

ITS ANCIENT SITES AND MODERN SCENES.

By GEORGE DENNIS,

Author of "The Cities and Cemeteries of Etruria."

With Illustrations. Post 8vo.

THE CORNWALLIS PAPERS;

THE PUBLIC AND PRIVATE CORRESPONDENCE OF CHARLES, MARQUIS CORNWALLIS, during the American War—his two Administrations in India—the Union with Ireland, and the Peace of Amiens. From Papers in Possession of the Family, and Official and other Documents, &c.

EDITED, WITH NOTES, BY CHARLES ROSS, ESQ.

THE WORK WILL INCLUDE LETTERS FROM

KING GEORGE III.	LORD MELVILLE.	LORD SIDMOUTH.
THE PRINCE OF WALES.	LORD CASTLEREAGH.	DUKE OF PORTLAND.
DUKE OF YORK.	LORD CORNWALLIS.	LORD LIVERPOOL.
	MR. PITT.	

Portrait. 3 vols. 8vo.

BRITISH INDIA.

A Summary of its Government, Resources, Races, and Religions.

CHAP. I.—SKETCH OF POLITICAL HISTORY, AND EXISTING GOVERNMENT—LAND TENURES—REVENUE SYSTEM—ORDERS IN COUNCIL—ACTS OF PARLIAMENT AND PARLIAMENTARY DOCUMENTS.
CHAP. II.—PROVINCES OF INDIA—COTTON—OPIUM—INDIGO—SALT—(PARLIAMENTARY PAPERS RELATING TO).
CHAP. III.—RACES AND RELIGIONS OF INDIA—HINDOO, MAHOMMEDAN, AND CHRISTIAN.

BY ARTHUR MILLS, M.P.

With Illustrative Maps. 8vo.

THE WORKS OF ALEXANDER POPE.

An entirely new edition: the Text carefully revised. Preceded by a Critical Essay on Pope and his former Editors. With more than 300 unpublished Letters.

EDITED BY THE LATE RIGHT HON. J. W. CROKER,

ASSISTED BY PETER CUNNINGHAM, ESQ.

8vo.

A LIFE OF ALEXANDER POPE.

To precede Mr. CROKER's Edition of the Works.

8vo.

BIOGRAPHICAL ESSAYS

CONTRIBUTED TO THE *EDINBURGH REVIEW* AND THE *QUARTERLY REVIEW*.

STRAFFORD.	STEELE.
CROMWELL.	CHURCHILL.
DE FOE.	FOOTE.

BY JOHN FORSTER.

8vo.

LIFE AND WORKS OF JONATHAN SWIFT, D.D.,

DEAN OF ST. PATRICK'S.

8vo.

SOME REMARKS ON GOTHIC ARCHITECTURE,

SECULAR AND DOMESTIC, PRESENT AND FUTURE.

BY GEORGE GILBERT SCOTT, Architect, A.R.A.

8vo.

WINGED WORDS ON CHANTREY'S WOODCOCKS.

EDITED BY JAMES P. MUIRHEAD, M.A.

With Etchings. Square 8vo.

MATERIALS TOWARDS A HISTORY OF AFGHANISTAN.

By J. P. FERRIER,
Adjutant-General of the Persian Army.

8vo.

ANCIENT POTTERY AND PORCELAIN:

Egyptian, Asiatic, Greek, Roman, Etruscan, and Celtic.

By SAMUEL BIRCH, F.S.A.

With many Woodcuts. 2 vols. Medium 8vo.

PRECEPTS FOR THE CONDUCT OF LIFE.

EXHORTATIONS TO A VIRTUOUS COURSE; DISSUASIONS FROM A VICIOUS CAREER. EXTRACTED FROM THE SCRIPTURES.

BY A LADY.

Fcap. 8vo.

THE STUDENT'S HUME.

A HISTORY of ENGLAND for the Upper Classes in Schools, based upon Hume's Work, incorporating the Researches of recent Historians, and continued down to the present time.

With Illustrations. Post 8vo.

Uniform with "The Student's Gibbon," "LIDDELL'S History of Rome," and "SMITH'S History of Greece."

"This work is designed to supply a long acknowledged want in our School Literature,—a History of England for Students, in a volume of moderate size, containing the results of the researches of the best modern historians, and free from sectarian and party prejudice. HUME'S 'HISTORY OF ENGLAND' has been chosen as the basis of the present work, on account of the excellence of his narrative, and the clearness of his style. Gibbon tells us, that the 'careless inimitable beauties of Hume often forced him to close the volume with a mixed sensation of delight and despair;' and it seemed worse than useless in a work like the present, to dispense with the assistance of so great a master in composition, and to repeat in other words a narrative that had been already so well told. But HUME has not always been followed as an *authority*, although he is more trustworty than most modern critics allow; still his language has been adopted as far as possible in the narration of events, while his facts and conclusions have been carefully sifted, especially in the History of the Constitution, and of political parties; and much new and important information has been added both from recent works and original documents."—*Extract from Preface.*

AN ENGLISH-LATIN DICTIONARY.

By WILLIAM SMITH, LL.D., and JOHN ROBSON, B.A.

8vo. and, Abridged, 12mo.

Uniform with Dr. SMITH'S Latin-English Dictionaries.

A MEDIÆVAL LATIN-ENGLISH DICTIONARY.

Selected and translated from the great work of DUCANGE.

8vo.

Uniform with "Dr. SMITH'S Latin Dictionary."

ESSAYS ON THE EARLY PERIOD OF THE FRENCH REVOLUTION.

By the late Rt. Hon. JOHN WILSON CROKER.

Reprinted from the "Quarterly Review." With additions, revised.

8vo.

A DICTIONARY OF BIBLICAL ANTIQUITIES,

By various Writers. Edited by WM. SMITH, LL.D.

Woodcuts. One Vol. Medium 8vo.

Uniform with the "Dictionary of Greek and Roman Antiquities."

"Dr. Smith's Dictionaries form an important element in our modern English scholarship. Probably no modern books have done so much to extend a knowledge of the researches and conclusions of the learned men of our time in the field of antiquity. If the Dictionaries to come are as well executed as their predecessors, the longer Dr. Smith continues to publish the better ordinary scholars will be pleased."—*Guardian.*

AN ATLAS OF ANCIENT GEOGRAPHY.

Forming a Companion Work to the "Dictionary of Greek and Roman Geography."

By WILLIAM SMITH, LL.D.

4to.

ENGLISH ROOTS AND RAMIFICATIONS:

Or, Explanations of the Derivation or Meaning of Divers Words in the English Language.

By JOHN ARTHUR KNAPP.

Fcap. 8vo.

Uniform with "TRENCH on Words," "HEAD's Shall and Will."

JOURNAL OF THE ROYAL GEOGRAPHICAL SOCIETY.
VOL. 27.

CONTENTS.

Anniversary Address on the Progress of Geography, Council Reports, &c.

1. JOCHMUS, GENERAL.—Expedition of Philip of Macedon against Thermus and Sparta. 2. Military Operations of Brennus and the Gauls against Thermopylæ and Ætolia. 3. Battle of Marathon. 4. Battle of Sellasia. Plans.
5. YULE, CAPT.—Geography of Burma and its Tributary States. Map.
6. MONTEITH, LT.-GEN.—Route from Bushir to Shiráz. Map.
7. LOFTUS, W. K.—The River "Eulæus" of the Greek Historians. Map.
8. OSBORN, CAPT. S.—The Sea of Azov, the Putrid Sea, &c. Map.
9. ABBOTT, CONSUL.—Notes on a Journey E. from Shiráz to Fessá and Dareb, thence W. by Kehrúm to Kazeran. Map.
10. RAWLINSON, SIR H. C.—Mohamrah and the Vicinity. Map.
11. GISBORNE, L.—Survey of the Isthmus of Darien. Map.
12. HOPKINS, THOS.—The Mild Winter-Temperature of the British Isles.
13. SPRATT, CAPT. T., R N.—Serpent Island.
14. CHAIX, PROF.—Hydrography of the Valley of the Arve.
15. CHEGHORN, J.—On the Water of Wick.
16. MACDONALD, J. D., Dr.—The Rewa River and its Tributaries. Map.
17. GRANT, LT.-COL.—Description of Vancouver Island.
18. ANDERSON, CHIEF FACTOR.—Route to Montreal Island.
19. LIVINGSTONE, Dr.—Routes in Central Africa. Map.
20. GREGORY, A. C.—North Australian Expedition. Map.

8vo.

New and Revised Editions.

HISTORY OF LATIN CHRISTIANITY,

Including that of the Popes to the Pontificate of Nicholas V.

By HENRY HART MILMAN, D.D., Dean of St. Paul's.

With an Index. A new Edition. 6 Vols. 8vo.

GREECE:

PICTORIAL, DESCRIPTIVE, AND HISTORICAL.

By CHRISTOPHER WORDSWORTH, D.D.,
Canon of Westminster.

With a History of the Characteristics of Greek Art.

By GEORGE SCHARF, F.S.A.

A new Edition, revised, with 600 Woodcuts. Royal 8vo.

ST. PAUL'S EPISTLES TO THE CORINTHIANS.

With Critical Notes and Dissertations.

By REV. ARTHUR P. STANLEY, M.A.,
Canon of Canterbury, and Regius Professor of Ecclesiastical History at Oxford.

A New and Revised Edition. 8vo.

ST. PAUL'S EPISTLES TO THE THESSALONIANS, GALATIANS, AND ROMANS.

WITH CRITICAL NOTES AND DISSERTATIONS.

By REV. B. JOWETT, M.A., Fellow and Tutor of Baliol College, Oxford.

Second and revised Edition. 2 Vols. 8vo.

SILURIA:

THE HISTORY OF THE OLDEST KNOWN ROCKS CONTAINING ORGANIC REMAINS; with a brief Sketch of the Distribution of Gold over the Earth.

By SIR RODERICK MURCHISON, D.C.L., F.R.S.,
Director General of the Geological Survey of the United Kingdom.

A new and thoroughly revised Edition, with Coloured Map, Plates, and Woodcuts. Medium 8vo.

THE CONNEXION OF THE PHYSICAL SCIENCES.

By MARY SOMERVILLE.

Ninth, and completely revised Edition. Woodcuts. Post 8vo.

PHYSICAL GEOGRAPHY.
By MARY SOMERVILLE.
Fourth, and completely revised Edition. Portrait. 1 Vol. Post 8vo.

A MANUAL OF SCIENTIFIC ENQUIRY,
Prepared for the Use of Officers of H.M. Navy on Foreign Service, and for Travellers in general.
By VARIOUS AUTHORS.
Third and revised Edition. Maps, &c. Post 8vo.
Published by Order of the Lords Commissioners of the Admiralty.

INSTRUCTIONS IN PRACTICAL SURVEYING,
PLAN DRAWING, AND SKETCHING GROUND WITHOUT INSTRUMENTS.
By G. D. BURR.
Third Edition. Plates. Post 8vo.

ELEMENTS OF GEOMETRY AND ALGEBRA.
For the Use of the Royal Hospital Schools, Greenwich.
By REV. GEORGE FISHER, M.A., Principal.
New and cheaper Editions. 16mo. 1s. 6d. each. (Ready.)

*** *The above Elementary Works are published by command of the Lords Commissioners of the Admiralty.*

HISTORICAL MEMORIALS OF CANTERBURY.
THE LANDING OF AUGUSTINE—THE MURDER OF BECKET—BECKET'S SHRINE— THE BLACK PRINCE.
By REV. A. P. STANLEY, M.A., Canon of Canterbury.
Fourth and Cheaper Edition. Woodcuts. Post 8vo. 7s. 6d. (Ready.)

A HISTORY OF INDIA:
THE HINDOO AND MAHOMMEDAN PERIODS.
By THE HON. MOUNTSTUART ELPHINSTONE.
Fourth Edition. Map. 8vo.

LIVES OF THE LINDSAYS;

Or, A MEMOIR OF THE HOUSES OF CRAWFORD AND BALCARRES.

By LORD LINDSAY.

Second and cheaper Edition. 3 Vols. 8vo.

"It is by no means a constant fact, that every heraldic painter, shall execute his labour of love and reverence with so much sincerity, delicacy, and patience, as Lord Lindsay has. He has given us a book which Scott would have delighted to honour."

"The critic's task would be a holiday labour—instead of being too often, as it is, a manufacture of bricks when the supply of straw again and again fails—if it led him more frequently to examine and exhibit such worthy books as Lord Lindsay's."—*Athenæum.*

LETTERS FROM HEAD-QUARTERS;

Or, THE REALITIES OF THE WAR IN THE CRIMEA.

By A STAFF OFFICER.

A new and condensed Edition. With Portrait and Plans. Post 8vo.

THE PURSUIT OF KNOWLEDGE UNDER DIFFICULTIES.

By G. L. CRAIK,

Professor of History and of English Literature in the Queen's University, Ireland.

A new Edition, with additional Examples. Portraits. 2 Vols. Post 8vo.

HOUSEHOLD SURGERY;

Or, HINTS ON EMERGENCIES.

By JOHN F. SOUTH, Surgeon to St. Thomas's Hospital.

A new Edition, revised, Chapters on Poisons and MEDICINE. Woodcuts. Post 8vo.

In this work useful hints are given as to the means which people have in their own power to employ when accidents happen which require immediate attention, and no medical man is at hand and often cannot be obtained for hours. Such cases are neither few nor unimportant, and many serious consequences, nay, even death, may be prevented, if a judicious person, having been put on the track, make use of the simple remedies which almost every house affords.

"We have seldom seen a book of wider or more sound practical utility than this unpretending little volume. We can conscientiously recommend Mr. South's Manual to the notice of the public. It is not a classbook; *it is everybody's book;* and above all, travellers, emigrants, and residents in remote country places should not fail to provide themselves with it."—*Morning Chronicle.*

SHALL AND WILL:

Or, TWO CHAPTERS ON FUTURE AUXILIARY VERBS.

By SIR EDMUND HEAD, BART., Governor-General of Canada.

New Edition. Fcap. 8vo.

Handbooks for Travellers.

HANDBOOK FOR INDIA.

Being an Account of the Three Presidencies, and of the Overland Route, and intended as a Guide for Travellers, Officers, and Civilians.

PART I.—MADRAS AND BENGAL.   PART II.—BOMBAY.

With Maps and Plans of Towns. Post 8vo.

HANDBOOK FOR SYRIA AND THE HOLY LAND.

Maps. Post 8vo.

HANDBOOK FOR SICILY.

Maps and Plans. Post 8vo.

HANDBOOK TO THE CATHEDRALS OF ENGLAND.

With Illustrations. Post 8vo.

HANDBOOK FOR PARIS.

Being a complete Guide for Visitors to all Objects of Interest in that Metropolis.

With Plans. Post 8vo.

HANDBOOK FOR KENT, SURREY, SUSSEX, AND HAMPSHIRE.

INCLUDING THE ISLE OF WIGHT.

Maps. Post 8vo.

HANDBOOK FOR TURKEY AND CONSTANTINOPLE.

A new and revised Edition. Maps. Post 8vo.

ALBEMARLE STREET,
October, 1857.

MR. MURRAY'S

LIST OF NEW WORKS.

Life of George Stephenson, the Railway Engineer.
By **SAMUEL SMILES**. Third Edition revised with additions. Portrait. 8vo. 16s.

"It is not too much to say, that Mr. Smiles has performed his office with eminent success, and a considerable void has been filled up in the page of modern history. We see the vast achievements and the epic story of this age of ours more than half comprised in the feats of its strongest and most successful worker. The worker himself, with his noble simplicity and energy, his zeal for his kind, his native born gentleness and indomitable tenacity, would probably have been eminent in any age or condition of society, but, in virtue of his actual achievements and the obstacles he surmounted, of his struggles and triumphs, we may designate him a hero, and ask in defence of this arbitrary title what real conditions of heroism there were wanting?"—*Times*.

Letters From High Latitudes.
Being some Account of a Yacht voyage to Iceland, Jan Mayen, and Spitzbergen, in 1856. By **LORD DUFFERIN**. Second Edition. Woodcuts. Crown 8vo. 21s.

"We should like extremely to go a yachting excursion with Lord Dufferin. His book is one which leads us not only to admire the talent and vivacity of its author, but to conceive a strong personal liking for him. In the most natural and unaffected way, he places his own picture before us. There is not a vestige of vapouring or boastfulness in the story—it is a quiet, manly statement of great dangers encountered like a true Briton. We see at once a high-spirited young man, always cheerful and good-natured, fond of fun, with an eye for the picturesque in scenery, and a taste for the romantic in history—determined to make light of hardships and inconveniences, and always to look at the bright side of things. The book is a most amusing and readable one."
Saturday Review.

Lives of Lords Kenyon, Ellenborough, & Tenterden.
Forming the Third and Concluding Volume of the "Lives of the Chief Justices of England." By **JOHN, LORD CAMPBELL**, LL.D., Chief Justice of England. With an Index to the entire Work. 8vo. 12s.

The Life and Opinions of General Sir Charles Napier;
chiefly derived from his Familiar Correspondence with his Family and Friends, and from his MSS. Journals. By **SIR WILLIAM NAPIER**, K.C.B. Second Edition. Portraits. 4 Vols. Post 8vo. 48s.

"Great men are soon forgotten, even if their greatness has ever been recognised, and Sir Charles Napier was so imperfectly understood during his life, that to nine-tenths of the readers of this Biography it will be matter of surprise that so little should be generally known of a man so wise, valiant, original, and noble, and so recently gone away from among us. Another is added to the long list of England's departed heroes—a man of whom yet unborn generations will be proud, and whose greatness was of a kind eminently fitted to find its way into the hearts of Englishmen.—*Saturday Review*.

A Residence Among the Chinese: INLAND, ON THE COAST, AND AT SEA.
Being a Narrative of Scenes and Adventures during a Third Visit to China, in 1853—56; with suggestions on the PRESENT WAR. By ROBERT FORTUNE. With Illustrations. 8vo. 15s.

"This book will be read with great interest. Manners, Scenery, Natural History, Commerce, Manufactures, and Agriculture,—especially as belongs to the two latter, the production of silk, and the cultivation of the tea shrub—and our attitude towards China at present, contribute, with many other subjects, to minister pleasure and instruction to the reader. No reader of this volume will think the time employed in its perusal to have been ill bestowed."
Christian Observer, Oct. 1857.

"Although this volume of Mr. Fortune's peregrinations in China does not substantially carry the reader over new ground, beyond a visit to the great silk district North of Shanghae, it is not the least interesting of the triad. The traveller exhibits a great mastery of his subject. He is more familiar with the country and the people. He travelled in his own character of foreigner, not in the guise of a native. He was consequently less fettered in his movements, and more at leisure to observe."
Guardian.

Pottery and Porcelain; Mediæval and Modern:
With Descriptions of the Manufacture in various Countries, the principal Collections, a Glossary, and a List of Monograms. By JOSEPH MARRYAT. Second Edition, Enlarged and Revised. Coloured Plates, Woodcuts, &c. Medium 8vo. 31s. 6d.

Memoirs left in MS. By the late Sir Robert Peel, Bart.
Edited by the Trustees of his Papers, EARL STANHOPE, and the RIGHT HON. EDWARD CARDWELL, M.P.

2 Vols. Post 8vo. 15s.

"The most simple, faithful, and valuable materials ever contributed to history. It is published just as the great statesman left it, and contains, in a complete form, the private and confidential narrative of the events connected with the carrying of Catholic emancipation. Indeed, it is less a continuous narrative than a collection of illustrative and explanatory documents; but these are so arranged as to form at once history and proof."—*Economist.*

Sinai and Palestine. In Connection with their History.
By REV. ARTHUR P. STANLEY, M.A. Regius Professor of Ecclesiastical History at Oxford, and Canon of Canterbury. Fourth Edition. Plans. 8vo. 16s.

"Very few English travellers have set out upon their pilgrimage with such advantage as Professor Stanley. The historical and critical works of this great master indicate the importance which he was early taught to attach to minute geographical detail as illustrating the historical records of the nations of antiquity, and the biographer of Dr. Arnold would be sure to regard as the most essential viaticum, a mind furnished with all the information requisite to turn his journey to the best account * * * Add to these qualifications his established reputation for the graces of style, and it is nothing strange that his 'Sinai and Palestine' should have attained a popularity which has been accorded to no book of Eastern travels since the publication of 'The Crescent and the Cross,' or 'Eothen'"—*Christian Remembrancer*, Oct., 1857.

Caravan Journeys and Wanderings in Persia, Herat,
AFGHANISTAN, AND BELOOCHISTAN. With Geographical and Historical notices of the countries lying between Russia and India. By J. P. FERRIER, Adjutant-General in the service of Persia. Second Edition. Map and Woodcuts. 8vo. 21s.

"Our quotations will do more to recommend the work to the general reader than any panegyric of our own. But we cannot help expressing our obligations to a writer, who has told a story, full of novel information and strange adventure, with so much modesty and intelligence."—*Edinburgh Review.*

Narrative of the Gunpowder Plot. By DAVID JARDINE, Barrister-at-Law. New Edition. Post 8vo. 7s. 6d.

"Mr. Jardine's 'Narrative' is not only the best book upon the subject which has yet appeared, but there is no probability of its ever being superseded by a better history of the same event."—*Notes and Queries.*

The History of the British Poor: in connection with the Condition of the People. By SIR GEORGE NICHOLLS, K.C.B., late Poor Law Commissioner, and Secretary to the Poor Law Board. 4 vols. 8vo.

"The conclusion of this work is, in fact, the conclusion of a History of the Poor Laws of Great Britain, which will be of essential service to all students of the progress of the country, and will assist greatly, no doubt, in prompting and directing future efforts for the perfecting of that 'charity in its largest application,' which its author as an active public servant has himself done so much to promote."—*Examiner.*

The Military Operations in Kaffraria, which led to the Termination of the Kaffir War; and on the Measures for the Protection and Welfare of the People of South Africa. By the late SIR G. CATHCART, K.C.B., Governor of the Cape of Good Hope. Second Edition, with Maps, 8vo. 12s.

Letters on Turkey; an Account of the Religious, Political, Social, and Commercial Condition of the Ottoman Empire, the Reformed Institutions, Army, Navy, &c. By M. A. UBICINI. 2 vols. Post 8vo. 21s.

"Ubicini has a deep knowledge of Turkish history and Mahometan literature, and a considerable acquaintance with the various races which inhabit Turkey; so that he brings a practical knowledge to correct or animate the written letter. In brief sketches he presents the pith of Turkish history on important epochs, and readily uses examples to illustrate his arguments.' He has carefully investigated the theory of Ottoman policy and practice; pointing out the real spirit of the Mahometan religion (much misunderstood), and its actual modifications; tracing the causes of Turkish decline, and stating the grounds of its probable regeneration."—*Spectator.*

Wanderings in Northern Africa, Benghazi, Cyrene, the Oasis of Siwah, &c. By JAMES HAMILTON. Woodcuts. Post 8vo. 12s.

"North Africa, although for many reasons so interesting, and particularly that division of it extending from the gulf of Sidra, or Sert, in the West, to Alexandria, Cairo, and the Nile, in the East (the route which Mr. James Hamilton chose for his travels), is to all general purposes, so little known to the English public, that in addition to its own intrinsic recommendations, the journey under notice claims attention on he ground nearly of absolute novelty. Mr. Hamilton brings to his work the acquirements of the scholar, and the penetration of the highly gifted literary artist."—*John Bull.*

Dictionary of Greek and Roman Geography. Edited by W. SMITH, LL.D. With Woodcuts. 2 Vols. Medium 8vo. 80s.

"As far as we have used this book (and it is only by constant use that the real worth of a Dictionary can be tested), we have never been disappointed."—*Guardian.*

The Historic Peerage of England. Exhibiting under alphabetical arrangement, the origin, descent, and present state of every Title of Peerage which has existed in this country since the Conquest. Being a new edition of "The Synopsis of the Peerage of England." By the late SIR HARRIS NICOLAS. Corrected to the present time by WILLIAM COURTHOPE, Somerset Herald. 8vo. 30s.

"In historic literature the new edition of Sir Harris Nicolas's 'Historic Peerage of England,' must take a high place. It exhibits 'under strictly alphabetical arrangement the descent of every title which has been conferred this country since the accession of William the Conqueror, the manner and period of its creation, the dates of the deaths of those who inherited it, and of the year when each dignity became extinct, was forfeited, or fell into abeyance.'"—*The Press.*

China: A General Description of that Empire and its Inhabitants,
with the History of Foreign Intercourse down to the Events which produced the Dissolution of 1857. By SIR JOHN F. DAVIS, BART. New Edition, revised and enlarged. Woodcuts. 2 Vols. Post 8vo. 14s.

"The publication of a new edition of Sir John Davis's book upon China is a natural result of the great interest in Chinese affairs which passing events have produced. The merits of the work itself are too well known to call for any very specific criticism on our part. We need only say that, in our opinion, it contains the most readable, and apparently the most credible account of the strange nation to which it refers."—*Saturday Review.*

Descriptive Essays: CONTRIBUTED TO THE QUARTERLY REVIEW
By SIR FRANCIS BOND HEAD, BART. 2 Vols. Post 8vo. 18s.

"Sir Francis Head writes only upon matters in the exposition of which he can make his personal experiences available; and the diversities of his experience are sufficiently remarkable. We trace him personally in South America, among the gorges of the Andes, and in the silver mines of Mexico; in the far west, amongst the tribes of the red man, whose habits he has studied; up the Rhine, with every valley and village of which he is familiar; and all over the well-known highways of Europe, by canal, river, horse-road, and rail. The hand of energetic utility is visible in all these articles."—*Literary Gazette.*

On some Disputed Questions of Ancient Geography.
By COL. W. MARTIN LEAKE, F.R.S. Map. 8vo. 6s. 6d.

"Col. Leake's investigations of classical antiquity are well known and highly valued. In his personal acquaintance with the places mentioned by ancient Greek and Roman writers, added to his accurate knowledge of the writers themselves, he possesses an advantage shared by few, if, any, other inquirers. Hence his opinions on all subjects connected with archæology are entitled to peculiar weight."—*Athenæum.*

Lives of the Lord Chancellors of England, from the
earliest times till the reign of George the Fourth. BY JOHN, LORD CAMPBELL, LL.D., Chief Justice of England. Fourth Edition, revised. With a carefully compiled Index. 10 Vols. Crown 8vo. 6s. each.

"We gladly welcome the work in this new and popular form, and think the learned and noble lord could hardly have bestowed a greater boon upon the profession of which he is so distinguished a member, than by placing so useful a book within the reach of all."—*Gent. Mag.*

The Early Flemish Painters; NOTICES OF THEIR LIVES
and Works. BY J. A. CROWE and G. B. CAVALCASELLE. With Woodcuts. Post 8vo. 12s.

"We welcome the work, as a diligent and painstaking monograph of the Schools of Bruges and Louvain, made doubly useful by its minute index. * * * A book which is all that we could wish, has most substantial merits, and cannot fail to be highly useful to the intelligent student of art."—*Christian Remembrancer, Oct. 1857.*

The Confidential Correspondence of Napoleon
BONAPARTE with his Brother Joseph, some time King of Spain. Selected and Translated from the French. 2 Vols. 8vo. 26s.

"This is a solid contribution to the history of the last generation. It is almost impossible to overrate the importance of such confidential outpourings of men who have themselves made history. The translator has rightly judged to let Napoleon speak for himself, and to give us no more comment than is absolutely necessary to make him understood. And we heartily thank him for his useful and judiciously employed labours."—*Guardian.*

The Principles of Surgery.
By **JAMES SYME**, F.R.S.E., Professor of Clinical Surgery in the University of Edinburgh. Fourth Edition. Revised. 8vo. 14s.

"A work on 'the Principles of Surgery' should, we think, be as remarkable for brevity as for perspicuity. We have never seen a work so massive in information, compressed within such narrow limits as the one now under consideration. It is worthy of especial note that the principles enunciated in previous editions remain unaltered in all essential particulars, while the volume itself, contrary to the received law of new editions, has actually in this last edition been to some extent shortened and condensed!"—*Scottish Press.*

The Art Treasures Exhibition at Manchester.
A WALK through the Building under the guidance of **DR. WAAGEN**. Post 8vo. 1s.

"Many persons will wish that Dr. Waagen's comment on the Manchester Exhibition had appeared earlier. We have now, in this companion to the catalogue, Dr. Waagen's remarks on the most important works of art. He goes the round of the building, and pronounces, as a connoisseur should, on their respective beauties."—*Guardian.*

An Atlas of the United States, Canada, New
BRUNSWICK, NOVA SCOTIA, NEWFOUNDLAND, MEXICO, CENTRAL AMERICA, CUBA, AND JAMAICA. From the most recent State Documents, Marine Surveys, and unpublished materials, with Plans of the principal Cities and Seaports, and an Introductory Essay. By **PROFESSOR ROGERS**, of Boston, U.S., and **A. KEITH JOHNSTON**, F.R.S.E. With 29 Plates. Folio, 63s.

*** *This is the only Collection of Maps of these Countries from documents not yet published in Europe or America.*

The Education of Character.
With Hints on Moral Training. By **MRS. ELLIS**, Author of "The Women of England," &c., &c. Post 8vo. 7s. 6d.

"This is a good book—a better book than we have seen from the same pen. It treats of education, private and public, higher and lower; of the causes which do so much to render it ineffective and even mischievous; and of the means by which it may be made more what it ought to be. The term 'Education' is used by Mrs. Ellis as embracing the culture of the moral character at least as much as the intellectual powers."—*British Quarterly Review, Oct., 1857.*

Sir William Blackstone's Commentaries on the
LAWS OF ENGLAND. A New Edition, adapted to the Present State of the Law. By **ROBERT MALCOLM KERR**, LL.D., Barrister-at-Law. 4 Vols. 8vo. 42s.

"The country gentleman's edition of Blackstone."—*Spectator.*

The Arts of the Middle Ages and Renaissance.
From the French of **M. JULES LABARTE**. With 200 Illustrations. 8vo. 18s.

"M. Labarte's knowledge is large, and he has the art of arranging it, with the systematic neatness of the French mind. The volume is illustrated with some of the most remarkable examples in every style of art. As a broad view of the domestic arts of the middle ages, and an introduction to their particular study, this 'Handbook' will be found extremely useful and satisfactory."—*Press.*

Insect Architecture.
To which are added Chapters on the Ravages, the Preservation for Purposes of Study, and the Classification of Insects. By **JAMES RENNE**, A.M. New Edition. Woodcuts. Post 8vo. 5s.

Later Biblical Researches in Palestine. Being a
Journal of Travels in the Year 1852. By REV. EDWARD ROBINSON, D.D. With Maps. 8vo. 15s.

"Persuaded as we are that no living writer has deserved so well of Sacred Geography as Dr. Robinson, the publication of a new volume of 'Biblical Researches' from his pen, was a subject of congratulation to all interested in Sacred Literature. Nor is its execution inferior to that of the earlier work to which the author owes his well-earned reputation."—*Christian Remembrancer, Oct. 1857.*

Five Years in Damascus. Including an account of the
History, Topography, and Antiquities of that City. With Travels and Researches in Palmyra, Lebanon, &c. By Rev. J. L. PORTER. With Map and Woodcuts. 2 Vols. Post 8vo. 21s.

"Mr. Porter's most valuable contribution to the geography of Southern Syria. His style is natural and easy, quite free from affectation; his descriptions always vigorous and life-like, sometimes even eloquent; his arguments and authorities well and forcibly stated and arranged, without parade and with perfect fairness. Mr. Porter must rank next to Dr. Robinson as a successful explorer of sacred lands."—*Christian Remembrancer, Oct., 1857.*

History of the Christian Church, FROM THE APOS-
TOLIC AGE TO THE CONCORDAT OF WORMS, A.D. 1122. By JAMES C. ROBERTSON, M.A., Vicar of Bekesbourne.

FIRST PERIOD.—TO THE PONTIFICATE OF GREGORY THE GREAT.
SECOND PERIOD.—TO THE CONCORDAT OF WORMS.

2 Vols. 8vo. 30s. May be had separately.

"The number of new Histories of the Church is a marked feature of the present day. We have them in all forms and sizes, and written by men of every degree of qualification, and of every division of the Christian Church. Mr. Robertson's is the best condensation we have met with. He is well read in the authorities; he quotes the originals with fearless honesty; and, although evidently a man of very decided opinions, he never intrudes them or writes up to them. He tells the story as he finds it without searching after novelty or striving to attract attention by paradox."—*Athenæum.*

History of Architecture IN ALL AGES AND ALL COUNTRIES.
By JAMES FERGUSSON. Third Thousand. With 850 Illustrations on Wood. 2 Vols. 8vo. 36s.

"A publication of no ordinary importance and interest. It fills up a void in our literature, which, with the hundreds of volumes we possess on that science, had never before been precisely attempted: and fills it up with learning and with ability."—*The Ecclesiologist.*

On the State of France before the Revolution,
1789; and on the causes that led to that event. By M. DE TOCQUEVILLE. Translated from the French by HENRY REEVE. 8vo. 14s.

"The appearance of this well-thought and well-reasoned work at this moment is most timely. M. de Tocqueville has spoken in a grave and earnest manner: mournfully, truthfully, and with the eloquence and ardour of the deepest conviction. His wise words will not sleep in the ears of his countrymen, and may in the fulness of time produce effects more lasting than many now suppose."—*Fraser's Magazine.*

Romany Rye: A Sequel to Lavengro. By GEORGE BORROW.
2nd. Edition. 2 vols. Post 8vo. 21s.

Lord Byron's Poetry Complete—Portable Edition. Printed
in a small but clear type, from the most correct text. Portrait and Index. Post 8vo. 9s.

"In compactness of size, and clearness and beauty of type, this is a model of a book for a TRAVELERS' LIBRARY. Mr. Murray's object has been to produce an edition of LORD BYRON, which should not encumber the portmanteau or carpet-bag of the Tourist. A more beautiful specimen of typography we have never seen."—*Notes and Queries.*

Aw. 202

Programming in Lua
fourth edition
_____

Programming in Lua
fourth edition

Roberto Ierusalimschy

Lua.org

Rio de Janeiro, 2016

Programming in Lua, fourth edition
by Roberto Ierusalimschy

ISBN 978-85-903798-6-7

Copyright © 2016, 2003 by Roberto Ierusalimschy. All rights reserved.

Published by the author. The author can be contacted at roberto@lua.org.

Book cover by Ana Ieru. Lua logo design by Alexandre Nako. Typeset by the author using LaTeX.

Although the author used his best efforts preparing this book, he assumes no responsibility for errors or omissions, or for any damage that may result from the use of the information presented here. All product names mentioned in this book are trademarks of their respective owners.

29 AP 58

Plate XI.

Lightning Source UK Ltd.
Milton Keynes UK
UKHW030630180422
401659UK00006B/841